Electron Scattering from Atoms, Ions and Molecules

Electron Scattering from Atoms, Ions and Molecules

Editors

Rajesh Srivastava
Dmitry V. Fursa

MDPI • Basel • Beijing • Wuhan • Barcelona • Belgrade • Manchester • Tokyo • Cluj • Tianjin

Editors

Rajesh Srivastava
Physics
Indian Institute of Technology
(IIT) Roorkee
Roorkee
India

Dmitry V. Fursa
Physics and Astronomy
Curtin University
Perth
Australia

Editorial Office
MDPI
St. Alban-Anlage 66
4052 Basel, Switzerland

This is a reprint of articles from the Special Issue published online in the open access journal *Atoms* (ISSN 2218-2004) (available at: www.mdpi.com/journal/atoms/special_issues/Electron_scatteringAtoms).

For citation purposes, cite each article independently as indicated on the article page online and as indicated below:

LastName, A.A.; LastName, B.B.; LastName, C.C. Article Title. *Journal Name* **Year**, *Volume Number*, Page Range.

ISBN 978-3-0365-7011-2 (Hbk)
ISBN 978-3-0365-7010-5 (PDF)

Contents

About the Editors

Rajesh Srivastava

Prof. Rajesh Srivastava is a distinguished theorist in atomic and molecular physics and works on the problems of atomic structures and various non-relativistic and relativistic collision processes as well as their applications to Plasma Physics. He is known for his work on electron–atom relativistic distorted wave (RDW) calculations. He completed his PhD in 1978 from the Banaras Hindu University (BHU) in India and thereafter completed his Postdoc from the FOM, the Institute of Atomic and Molecular Physics in Amsterdam. He has been working at the Indian Institute of Technology (IIT) Roorkee since 1983 and has over 40 years of teaching and research experience. He has published over 215 research publications in international journals, nearly 240 papers in conferences, and supervised many PhD students. He has delivered several invited talks and has been a Visiting Professor/Fellow at various institutions, viz., in the USA, Holland, the UK, Australia, Austria, Japan, Canada, and Germany.

Dmitry V. Fursa

Prof. Fursa works in the field of the computational modelling of quantum reaction processes. He received his PhD in 1995 from the Flinders University of South Australia. He has been a Fellow of the American Physical Society since 2012. Prof. Fursa has worked at Curtin University since 2007, where he conducts an active research program, teaches several undergraduate units, and supervises undergraduate and PhD students. The research program aims to model the interactions of light particles (electrons, positrons, and photons) with atoms, ions, and molecules. Emphasis is put on providing comprehensive collision datasets that are widely used in fusion research, astrophysics, health science, and the lighting industry.

Preface to "Electron Scattering from Atoms, Ions and Molecules"

Electron collision physics covers a broad range of processes in atoms and molecules. Understanding these processes can be achieved via experimental and theoretical investigations that support and challenge each other. The last few decades have been witness to tremendous progress in both the computational and experimental techniques applied to study and model electron-driven processes. Access to modern supercomputer facilities has allowed for the computational modelling of collision processes involving complex atoms and molecules; in turn, this allows for a sophisticated modelling and the diagnostic assessment of various plasmas. Applications of electron collision physics range from fusion, precision measurement and attoclocks to radiation damage and biomedical research. This volume collates diverse applications of collision physics, highlighting the importance and power of theoretical and computational techniques while also presenting new experiments which disclose exciting new developments in collision processes.

Rajesh Srivastava and Dmitry V. Fursa
Editors

atoms

Editorial

"Atoms" Special Issue (Electron Scattering from Atoms, Ions and Molecules)

Rajesh Srivastava [1,*] and Dmitry V. Fursa [2]

1 Department of Physics, Indian Institute of Technology (IIT), Roorkee 247667, India
2 Department of Physics and Astronomy, Curtin University, Perth, WA 6845, Australia
* Correspondence: rajesh.srivastava@ph.iitr.ac.in

Citation: Srivastava, R.; Fursa, D.V. "Atoms" Special Issue (Electron Scattering from Atoms, Ions and Molecules). *Atoms* **2023**, *11*, 31. https://doi.org/10.3390/atoms11020031

Received: 3 February 2023
Accepted: 3 February 2023
Published: 6 February 2023

Electron collision physics covers a broad range of processes in atoms and molecules. Understanding these processes can be achieved via experimental and theoretical investigations that support and challenge each other. The last few decades have seen tremendous progress in both the computational and experimental techniques applied to study and model electron-driven processes. Access to modern supercomputer facilities has allowed for the computational modelling of collision processes involving complex atoms and molecules [1–6] and this in turn allows for sophisticated modelling and diagnostic assessment of various plasmas [1,7,8]. Applications of electron collision physics range from fusion [3,7], precision measurement and attoclocks [9] to radiation damage and biomedical research [10,11]. This volume collates diverse applications of collision physics, highlighting the importance and power of theoretical and computational techniques while also presenting new experiments which disclose the exciting new developments in collision processes.

Pulsed mater injection in plasma devices is an actively growing research direction with many important applications including fusion, material processing, plasma–surface interactions, plasma thrusters, etc. Sadek et al. in [7] report on experiments with a magneton RF plasma which operated in argon. Pulsed argon gas injection is analyzed by optical emission spectroscopy of argon 2p-1s transitions. The measured line intensities in the 700–900 nm wavelength range are compared with those computed from a collisional–radiative (CR) model in order to determine time-resolved electron temperature in the pulsed injection conditions. In the CR modelling, the population mechanisms under consideration are direct and stepwise electron impact excitation (including cascades from high-energy levels), the while depopulation mechanisms considered are radiative transitions (mitigated by radiation trapping) and quenching reactions induced by collisions with neutral argon atoms. It was demonstrated that the full CR model better describes the optical spectrum than the coronal model, thereby revealing the importance of considering mechanisms involving Ar 1s levels and especially radiation trapping.

Plasma diagnosis via the utilization of optical emission spectroscopy (OES) is a powerful technique used to study plasma kinetics and gain knowledge of the production rate of different species present in the plasma. Information on key plasma parameters, such as electron temperature and electron density, can be obtained by comparing the OES measurements with the results of plasma modelling. To assess plasmas that significantly deviate from the equilibrium conditions, appropriate collisional–radiative (CR) models must be developed. Shukla et al. in [1] report on the development of a such model for Ar and Ar^+ plasma. Their CR model uses a comprehensive set of cross sections to describe electron impact excitations between different fine-structure resolved levels of Ar and Ar^+. The cross sections are obtained using relativistic distorted wave theory and are generally in good agreement with more accurate but more restrictive R-matrix calculations. In addition, the electron impact ionization, radiation trapping, diffusion, and three-body recombination processes are considered. Argon plasmas are of particular interest due to a considerable discrepancy between the electron temperature values obtained from the Langmuir probe and the line-ratio estimates in argon helicon plasma. The CR model was applied to an argon helicon plasma reported by Soltani and Habibi [12] for both ICP and helicon modes at various powers.

The interaction of high-energy radiation with condensed matter leads to the production of a large number low-energy electrons and ions via a cascade of ionization processes. In biological matter, these secondary electrons can interact resonantly or directly with biomolecules, causing damage to the DNA and the RNA in terms of either single or double-strand breaks. The complete set of absolute cross sections resulting from low- to intermediate-energy electron collisions with DNA molecules is required for radiation damage modelling, e.g., in Monte Carlo particle track simulation. Such cross sections include the total ionization and dissociative ionization cross sections. However, the cross sections for dissociative ionization, also known as partial cross sections, are very scarce. Rehman and Krishnakumar in [10] have used a crossed beam electron–molecular experiment along with the relative flow technique to measure absolute total and partial ionization cross sections of adenine molecules. The most abundant fragment cations from adenine include $C_nH_nN_n^+$ (n = 5, 4, 3, 2, 1) at m/z of 135 ($C_5N_5H_5$+), 108 ($C_4N_4H_4^+$), 81 ($C_3N_3H_3^+$), 54 ($C_2N_2H_2^+$), 27 (CNH^+), and $HCNH^+$. Good agreement was found with binary-encounter dipole calculations of Huo et al. [13]. The dominance of $C_nH_nN_n^+$ (n = 1 to 5), confirmed by the cross section measurements, suggests that the most favored pathway for adenine dissociation via electron ionization is effective due to the loss of HCN molecules in succession.

Dissociative electron attachment (DEA) is the dominant pathway for the interaction of low-energy electrons with molecules. DEA is a resonant process whereby the electron energy is translated to the nuclear motion via the dynamics of the negative ion resonance state (NIRA). A fascinating feature of DEA is site selectivity, which directly correlates with the functional group present at the site and originates from the nature of NIRS as core excited resonances. Tadsare et al. in [11] studied DEA processes in aromatic molecules aniline and benzylamine. These are present in many biological molecules, including DNA bases. It was found that H^- and CN^- are two dominant channels in the measurements of DEA to aniline and benzylamine, with H^- being the most dominant for both molecules. The absolute cross sections, as a function of electron energy for these channels, have been produced from both molecules. The DEA dynamics of the H^- channel has been investigated using the velocity slice imaging technique. The kinetic energy and angular distribution of hydride anions, formed in DEA to aniline and benzylamine, have been determined. The results of the investigation show that the functional group-dependent site-selective fragmentation, observed in aliphatic compounds, can also be seen in aromatic compounds.

In most of the electron–atom elastic scattering studies, a model potential approach is used consisting of static, exchange, polarization and absorption potentials [14]. The model potential is then used in the Schrodinger or Dirac equations depending upon whether the calculation is non-relativistic or relativistic. The equations can be solved through partial wave phase shift analysis and scattering phase shifts are obtained. The cross sections are eventually calculated using the phase shifts. In order to calculate model potential, charge density of the atom is required to obtain the static and exchange potentials and polarizability of the atom for calculating the polarization potential. Thus, accurate atomic structure calculations must be performed to establish the charge density and polarizability in addition to solving the scattering problem. Sahoo [2], using relativistic coupled-cluster (RCC) theory [15], via his calculations provided the electron density functions for obtaining the static and exchange potentials of the atoms. He presented the accurate electron densities and electric polarizabilities of Be, Mg, Ne and Ar atoms using two variants of the RCC method. Using these quantities, model potentials for the electron scattering of these atoms can be constructed. He also evaluated the second- and third-order electric dipole and quadrupole polarizabilities using a linear response approach.

Electron scattering cross sections of neutral tin are important for fusion research. Indeed, they have applications as fusion reactors, e.g., ITER. Accurate and comprehensive collision data for electron scattering on both neutral and all ion stages of tin will enable the modelling of plasma containing tin and the identification of tin spectral signatures across the different regions of the fusion plasma [16]. The collision data available for electron scattering on atomic tin are limited to several theoretical works and one experimental study. In view of this, Umer et al. [3] carried out electron scattering cross sections calculations from atomic tin using the relativistic convergent close-coupling method [17]. They presented integrated and momentum transfer cross sections for elastic scattering from the ground and the first four excited states of tin. Various integrated and selected differential cross sections are presented for excitation of the $5p^2$, $5p6s$, $5p5d$ and $5p6p$ manifolds from the ground state. The total ionization cross sections are calculated from the ground and the first four excited states, accounting for the direct ionization of the $5p$ valence shell and the closed 5s shell and the indirect contributions from excitation–autoionization.

In the scattering calculations from a projectile on a target, the results are reported in terms of different cross sections. However, the scattering time delay has garnered recent attention because it allows for the characterization of the projectile–target interaction in the temporal domain. In general, it is assumed that scattering interactions constitute instantaneous responses to the incident projectile. In reality, attoseconds (as) of delay in the time scale are needed to determine the specific interaction. This means that, compared to the projectile that does not feel the scattering centre, the scattered particles experience a time lag or time advancement. This phenomenon can be represented through a parameter known as Eisenbud–Wigner–Smith (EWS) time delay [18]. Aiswarya and Jose [4] presented a study on the angular time delay of e-C_{60} elastic scattering in their paper. They employed the annular square well (ASW) potential to simulate the C_{60} environment. In fact, the time delay in electron scattering depends on both the scattering angle θ and scattered electron energy E. The contribution from different partial waves to the total angular time delay profile was examined in detail.

There are various ions which are considered for carrying out high-precision measurements such as testing Lorentz symmetry violations, parity nonconservation effects, non-linear isotope shift effects and quantum information, including for the optical atomic clock experiments. In fact, optical lattices, when blended with unique features of optical transitions in the ionic system, lead to the revolution in the clock frequency states. Singly charged alkaline-earth ions are the most eligible candidates for consideration for use high-precision measurements due to several advantages they possess. Since the confinement of Mg^+ ions in a monochromatic optical dipole trap has become feasible experimentally for several ms, the pathways to implement these ions and thus realize optical lattice clocks have been opened up due to the fact that ions provide more accurate atomic clocks. This is due to the fact that various systematics in the ions can be controlled easily. Jyoti et al. [9] reported the magic wavelengths (λ_{magic}s) [19] and tune-out wavelengths (λ_Ts) of many $S_{1/2}$ and $D_{3/2,5/2}$ states, as well as transitions among these states of the Mg^+, Ca^+, Sr^+ and Ba^+ ions that are independent of M values. In many earlier studies, values of the electric dipole (E1) matrix elements were inferred precisely by combining measurements and calculations of λmagic. Similarly, the λ_T values of an atomic state can be used to infer the E1 matrix element [20].

Arretche et al. [5] studied the low-energy scattering of electron–Zn/Cd by applying model exchange and semiempirical polarization potentials. Their study was motivated by the fact that the total cross section measurements for electron scattering by Zn and Cd performed previously and the existence of p-wave shape resonances below 1 eV are well established in the literature. It was suggested that a second d-wave shape resonance could exist in both systems at an energy slightly higher than the one recorded for the p-wave but still below the inelastic threshold. In view of this, they reported the elastic scattering calculations for electron collisions with Zn and Cd atoms below 4 eV using a semiempirical approach [21], as well the scattering length for both targets. Their results show that the d-wave shape resonance can be found in Zn but is absent in Cd.

Campbell et al. [8] reported an interesting study on the inclusion of electron interactions by rate equations in chemical models [22]. The concept of treating subranges of the electron energy spectrum as species in chemical models was investigated. This is intended to facilitate the simple modification of chemical models by incorporating the electron interactions as additional rate equations. It is anticipated that this embedding of the fine details of the energy dependence of the electron interactions into rate equations will yield an improvement in computational efficiency compared to other methods. In their study, the authors proposed and tested a method to simulate nonequilibrium interactions of electrons with gas molecules. In this method, the energy range of the electrons is split into subranges that are then treated in a time-step calculation in the same way as chemical species. As such, the electron interactions can be incorporated easily into existing simulations without new coding being required. The authors found that, in excitation of gas molecules with one vibrationally excited level, the initial energy of the electrons was transferred to the gas molecules until an equilibrium was reached that, with sufficiently small subranges, was very close to the predicted equilibrium values. This equilibrium was then maintained over a long time (10^6 s), validating the method of calculating the rates for the electron interactions. It was observed that the simulated electron spectrum was also very close to the predicted Maxwellian distribution. Thus, they concluded that their proposed method is capable of producing accurate results. However, the minimum number of subranges, and thus computational efficiency, will need to be assessed for the requirements and circumstances of particular applications.

In recent years, the electron impact ionization of atoms, molecules and ions which are the most fundamental atomic processes has been studied using different theoretical and experimental techniques. Electron impact ionization, also referred to as (e, 2e), involves the collision of an incident electron with a target (either an atom or an ion or a molecule), leading to the ionization of the target [23]. Upon determining the energies and the momenta of all the particles involved in the collision, complete understanding of the ionization process is established. Thus, (e, 2e) collisions have become an important tool for investigating the collision dynamics of targets. The triple differential cross section (TDCS) is the physical quantity that is of prime interest in these studies, providing information about collision processes, ionization mechanisms, and the dependence of the ionization process on the electron kinematics under which ionization is taking place. Pandey and Purohit [6] carried out calculations for the electron impact triple differential cross section (TDCS) and reported the results for nitrogen molecules. The TDCSs have been obtained using distorted wave Born formalism [24] by taking the orientation averaged molecular orbital (OAMO) approximation.

This current volume of the Special Issue presents a collection of interesting papers related to electron scattering from various atoms and molecules and explores the possible applications of these studies and their findings to plasma physics.

Author Contributions: All authors contributed equally. All authors have read and agreed to the published version of the manuscript.

Conflicts of Interest: The authors declare no conflict of interest.

References

1. Shukla, N.; Gangwar, R.K.; Srivastava, R. Diagnostics of Argon Plasma Using Reliable Electron-Impact Excitation Cross Sections of Ar and Ar+. *Atoms* **2022**, *10*, 118. [CrossRef]
2. Sahoo, B.K. Constructing Electron-Atom Elastic Scattering Potentials Using Relativistic Coupled-Cluster Theory: A Few Case Studies. *Atoms* **2022**, *10*, 88. [CrossRef]
3. Umer, H.; Bray, I.; Fursa, D.V. Cross Sections for Electron Scattering from Atomic Tin. *Atoms* **2022**, *10*, 78. [CrossRef]
4. Jose, J. An Investigation of the Resonant and Non-Resonant Angular Time Delay of e-C60 Elastic Scattering. *Atoms* **2022**, *10*, 77.
5. Arretche, F.; Tenfen, W.; Sahoo, B.K. Semiempirical Calculations on Low-Energy Electron Scattering by Zn and Cd Atoms. *Atoms* **2022**, *10*, 69. [CrossRef]
6. Pandey, A.; Purohit, G. TDCS Calculation for the Ionization of Nitrogen Molecule by Electron Impact. *Atoms* **2022**, *10*, 50. [CrossRef]
7. Sadek, T.; Vinchon, P.; Durocher-Jean, A.; Carnide, G.; Kahn, M.L.; Clergereaux, R.; Stafford, L. Time-Resolved Analysis of the Electron Temperature in RF Magnetron Discharges with a Pulsed Gas Injection. *Atoms* **2022**, *10*, 147. [CrossRef]

8. Campbell, L.; Muccignat, D.L.; Brunger, M.J. Inclusion of Electron Interactions by Rate Equations in Chemical Models. *Atoms* **2022**, *10*, 62. [CrossRef]
9. Jyoti; Kaur, H.; Arora, B.; Sahoo, B.K. Magnetic Sublevel Independent Magic and Tune-Out Wavelengths of the Alkaline-Earth Ions. *Atoms* **2022**, *10*, 72. [CrossRef]
10. Rehman, M.A.; Krishnakumar, E. Electron Impact Ionization of Adenine: Partial Cross Sections. *Atoms* **2022**, *10*, 100. [CrossRef]
11. Tadsare, V.; Das, S.; Gokhale, S.; Krishnakumar, E.; Prabhudesai, V.S.S. Dynamics of Site Selectivity in Dissociative Electron Attachment in Aromatic Molecules. *Atoms* **2022**, *10*, 98. [CrossRef]
12. Soltani, B.; Habibi, M. Development of a Helicon Plasma Source for Neutral Beam Injection System of the Alborz Tokamak. *J. Fusion Energy* **2017**, *36*, 152. [CrossRef]
13. Huo, W.M.; Dateo, C.E.; Fletcher, G.D. Molecular data for a biochemical model of DNA damage: Electron impact ionization and dissociative ionization cross sections of DNA bases and sugar-phosphate backbone. *Radiat. Meas.* **2006**, *41*, 1202–1208. [CrossRef]
14. Mahato, D.; Sharma, L.; Srivastava, R. A new approach to study electron and positron scattering from acetylene. *J. Electron Spectros. Relat. Phenom.* **2021**, *252*, 147118. [CrossRef]
15. Chakraborty, A.; Rithvik, S.K.; Sahoo, B.K. Relativistic normal coupled-cluster theory analysis of second- and third-order electric polarizabilities of Zn I. *Phys. Rev. A* **2022**, *105*, 062815. [CrossRef]
16. Sharma, L.; Bharti, S.; Srivastava, R. Electron impact excitation of tin. *Eur. Phys. J. D* **2017**, *71*, 121. [CrossRef]
17. Fursa, D.V.; Bray, I. Fully Relativistic Convergent Close-Coupling Method for Excitation and Ionization Processes in Electron Collisions with Atoms and Ions. *Phys. Rev. Lett.* **2008**, *100*, 113201. [CrossRef]
18. Smith, F.T. Lifetime Matrix in Collision Theory. *Phys. Rev.* **1960**, *118*, 349. [CrossRef]
19. Liu, P.L.; Huang, Y.; Bian, W.; Shao, H.; Guan, H.; Tang, Y.B.; Li, C.B.; Mitroy, J.; Gao, K.L. Measurement of Magic Wavelengths for the $^{40}Ca^+$ Clock Transition. *Phys. Rev. Lett.* **2015**, *114*, 223001. [CrossRef]
20. Sahoo, B.K.; Wansbeek, L.W.; Jungmann, K.; Timmermans, R.G.E. Light shifts and electric dipole matrix elements in Ba^+ and Ra^+. *Phys. Rev. A* **2009**, *79*, 052512. [CrossRef]
21. Arretche, F.; Barp, M.V.; Scheidt, A.; Seidel, E.P.; Tenfen, W. Semiempirical models for low energy positron scattering by Ar, Kr and Xe. *J. Phys. B At. Mol. Opt. Phys.* **2019**, *52*, 215201. [CrossRef]
22. Kim, Y.H.; Fox, J.L. The Chemistry of Hydrocarbon Ions in the Jovian Ionosphere. *Icarus* **1994**, *112*, 310–325. [CrossRef]
23. Lahmam-Bennani, A. Recent developments and new trends in (e,2e) and (e,3e) studies. *J. Phys. B At. Mol. Opt. Phys.* **1991**, *24*, 2401. [CrossRef]
24. Madison, D.H.; Al-Hagan, O. The Distorted-Wave Born Approach for Calculating Electron-Impact Ionization of Molecules. *J. At. Mol. Phys.* **2010**, *2010*, 367180. [CrossRef]

atoms

MDPI

Article

Time-Resolved Analysis of the Electron Temperature in RF Magnetron Discharges with a Pulsed Gas Injection

Thibault Sadek [1,2], Pierre Vinchon [1], Antoine Durocher-Jean [1], Guillaume Carnide [2,3], Myrtil L. Kahn [3], Richard Clergereaux [2] and Luc Stafford [1,*]

1 Département de Physique, Université de Montréal, Montreal, QC H2V 0B3, Canada
2 LAPLACE (Laboratoire Plasma et Conversion d'Energie), Université de Toulouse, CNRS, 31000 Toulouse, France
3 LCC (Laboratoire de Chimie de Coordination), CNRS, 31000 Toulouse, France
* Correspondence: luc.stafford@umontreal.ca

Abstract: Pulsed gas injection in a plasma can affect many fundamentals, including electron heating and losses. The case of an asymmetric RF magnetron plasma with a pulsed argon injection is analyzed by optical emission spectroscopy of argon 2p-to-1s transitions coupled with collisional-radiative modeling. For a fully detailed population model of argon 2p levels accounting for direct and stepwise electron-impact excitation in optically thick conditions, a rapid decrease in the electron temperature, T_e, is observed during each gas injection with the sudden pressure rise. The opposite trend, with unrealistic T_e values before and after each pulse, is observed for analysis based on simple corona models, thus emphasizing the importance of stepwise excitation processes and radiation trapping. Time-resolved electron temperature variations are directly linked to the operating parameters of the pulsed gas injection, in particular the injection frequency. Based on the complete set of data, it is shown that the instantaneous electron temperature monotonously decreases with increasing pressure, with values consistent with those expected for plasmas in which charged species are produced by electron-impact ionization of ground state argon atoms and lost by diffusion and recombination on plasma reactor walls.

Keywords: optical emission spectroscopy; collisional radiative modeling; RF plasma; magnetron discharges; pulsed gas injection

Citation: Sadek, T.; Vinchon, P.; Durocher-Jean, A.; Carnide, G.; Kahn, M.L.; Clergereaux, R.; Stafford, L. Time-Resolved Analysis of the Electron Temperature in RF Magnetron Discharges with a Pulsed Gas Injection. *Atoms* **2022**, *10*, 147. https://doi.org/10.3390/atoms10040147

Academic Editors: Rajesh Srivastava and Dmitry V. Fursa

Received: 10 October 2022
Accepted: 29 November 2022
Published: 6 December 2022

1. Introduction

Pulsed matter injection is an appealing method in many plasma devices. It consists of the pulsed injection of gases, liquids, aerosols, or sprays. This can be done by gas puffing or supersonic flows [1,2], by gas or liquid valves, by ink-jet printer heads [3–5] or by direct liquid injection [6–8]. Pulsed gas injection enables to control the discharge physics. As an example, long term discharges in fusion devices can be sustained by the pulsed injection of fuel gas [9,10]. For supersonic gas injection, a significant increase in the fueling efficiency is observed due to the short injection time and the prompt cooling of the plasma edge consecutive to the massive injection of matter [9]. In RF discharges used for sputtering applications, pulsed injection of N_2 in Ne can induce a transition between two excitation mechanisms of plasma particles, from Penning reaction to electron-impact excitation [11,12]. Pulsed gas injection also enables to control the gas fluid dynamics. For example, a continuous injection of gases in a given reactor can produce an asymmetric flow and, consequently, an asymmetric distribution of the plasma parameters that pulsed plasma operation can avoid. Indeed, during the afterglow between pulses, species can efficiently diffuse in the plasma volume yielding to uniform power deposition [13]. The gas flow can also be efficiently guided to regions of interest, for example near a substrate or a target [14].

Pulsed gas injection further enables to control the plasma chemistry. First, depending on the operating conditions (gas pressure, reactor dimensions, power, etc.), it can affect the reaction kinetics involved in the plasma volume and at the plasma reactors walls [15]. For example, the production of metal-carbon clusters of higher masses is improved with the pulsed injection of He/CH_4 gas mixtures in arc discharges [16]. In addition, a pulsed injection of organosilicon precursors in an asymmetric RF plasma allows to combine plasma-enhanced chemical vapor deposition and physical vapor deposition without target poisoning [17]. Such conditions can further induce nucleation of nanoparticles in the plasma volume [18]. It is also possible to design the region of reactive species production. For example, a selective production of the precursors and deposition can be achieved by isolating spatially and temporally the injected gases from the plasma zone using a pulsed gas injection coupled with a pulsed injection of power [13,19]. In addition, many reactive precursors can be introduced in a pulsed mode. It allows to manipulate complex precursors such as liquids, metalorganic compounds diluted in solvents [20], as well as to introduce liquid dispersions of nanoparticles [8]. The latter is an attractive method for the formation of nanocomposite coatings based on at least one nanometer-sized (less than 100 nm) phase. The properties of such materials can be tailored by the chemical composition, crystal structure, and morphology of the matrix, but also by the nature, size, form, volume fraction, and distance between each nanometric inclusions (particles, filaments, tubes) [21,22]. Many studies are devoted to plasma processes with pulsed gas injection [23]. A new "hybrid" method based on a pulsed injection mode has been recently proposed [24]. It consists of a Direct Liquid Reactor-Injector (DLRI) in which nanoparticles are synthesized by mixing a liquid and a gaseous precursor prior to their injection in a pulsed mode in the plasma. The nanoparticles as well as the solvent are then injected in the plasma as a gas pulse with a duration and a frequency set by the chemical reaction kinetics.

In contrast to usual plasmas operating under constant pressure, the conception of advanced plasma processes with a pulsed gas injection inevitably implies a complex temporal dynamic associated with sudden pressure and plasma phase composition variations. Such feature can induce multiscale variations of the fundamental plasma properties, including electron density and temperature, number density of excited species, neutral gas temperature, etc. [18,25]. During nanocomposite thin film deposition using a DLRI, this can play an important role (i) on the dissociation kinetics of the matrix precursor [26], (ii) on the charging and transport dynamics of nanoparticles in the plasma [27], (iii) on the plasma-substrate interaction during thin film deposition [28], and therefore, (iv) on the physical and chemical properties of the coatings [29]. As a building block towards a better understanding of plasma processes using the DLRI, the objective of this study is to gain insights into the physics driving low-pressure RF plasmas with pulsed gas injection. Experiments are done in the specific case of a magnetron RF plasma operated in argon [30]. Optical emission spectroscopy combined with collisional-radiative modelling of Ar 2p-to-1s transitions is used to analyze the electron kinetics over a wide range of DLRI conditions.

2. Experimental Setup

2.1. Plasma Reactor and Pulsed Gas Injection

A schematic of the RF (13.56 MHz) magnetron plasma reactor used in this work is presented in Figure 1. The system consists of a 28 cm × 28 cm (diameter × height) cylindrical stainless-steel chamber with grounded walls. A two-stage pumping system ensures a residual vacuum of 10^{-5} Torr. Argon is fed into the reactor using an Atokit from Kemstream® plugged on the injection ring surrounding the top electrode. The injection parameters for the set of experiments reported in this work are the opening time of the first valve before the mixing chamber (5 ms), the opening time of the outer valve leading to the injection ring (10 ms), and the offset time between the closure of the two valves (2 ms). In these experiments, the frequency of the Ar injection pulses is varied and sets sequentially at f = 0.1, 0.5 and 1 Hz. Pulsed gas injection is producing an overpressure, Δp, in the plasma chamber. Values of Δp are mainly controlled by the injection parameters and the base pressure, p_0. The latter is adjusted using a throttle valve located at the entrance of the pumping system.

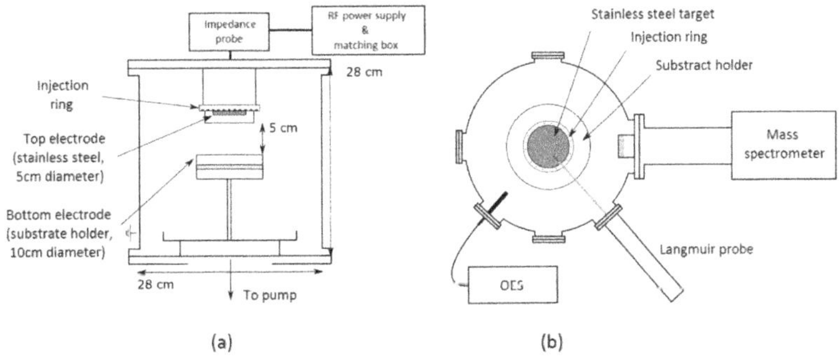

Figure 1. Overview of the experimental set up from (**a**) a side view and (**b**) a top view.

The RF power is fed to the top stainless-steel electrode (5 cm diameter), while the bottom electrode (10 cm diameter) is grounded. This results in an asymmetrical RF plasma with an inter-electrode gap of 5 cm. A matching box set between the RF power supply and the reactor is used to minimize the reflected power. For the experiments reported in this work, a nominal power of 100 W is sent to the plasma and the reflected power is oscillating with the gas pulses between 3 and 7 W.

2.2. Optical Emission Spectroscopy Measurements

As shown in Figure 1b, optical emission spectroscopy (OES) measurements are taken from a port and a collimator located on the side of the reactor. They are recorded over the 700–900 nm wavelength range with an AVANTES spectrometer (AVASpec-3648-2-USB) having a spectral resolution of ~0.16 nm (full width at half maximum). A typical spectrum depicting most of the intense Ar 2p-to-1s transitions (corrected for both background noise and apparatus spectral response) from a plasma generated at 5 mTorr and 100 W is shown in Figure 2.

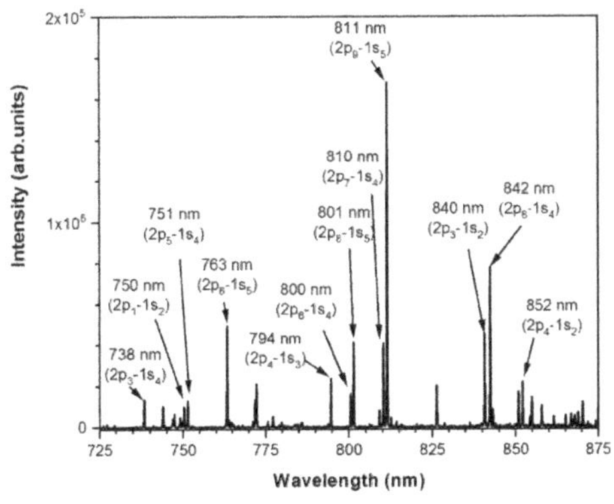

Figure 2. Typical optical emission spectrum of the argon plasma examined in this work.

For all conditions reported in this work, the measured line intensities are compared to those computed from a collisional-radiative model based on Donnelly's trace-rare-gases optical emission spectroscopy method [31] in order to determine time-resolved electron temperature in these pulsed injection conditions. As discussed previously [18,32,33], the model is however adapted with respect to Donnelly's work to account for radiation trapping and stepwise excitation through resonant 1s levels.

3. Results and Discussion

3.1. Influence of the Pressure Pulses

Optical emission spectra are recorded along a pulse of gas injection for an injection frequency of 0.1 Hz. Figure 3 shows the evolutions of the operating pressure (blue plot) and the Ar 811 nm emission line intensity during a period of injection (red plot) and for two conditions of base pressure set to p_0 ~5 mTorr (a) and 160 mTorr (b)—the constant pressure rise being of Δp ~70 mTorr during each cycle.

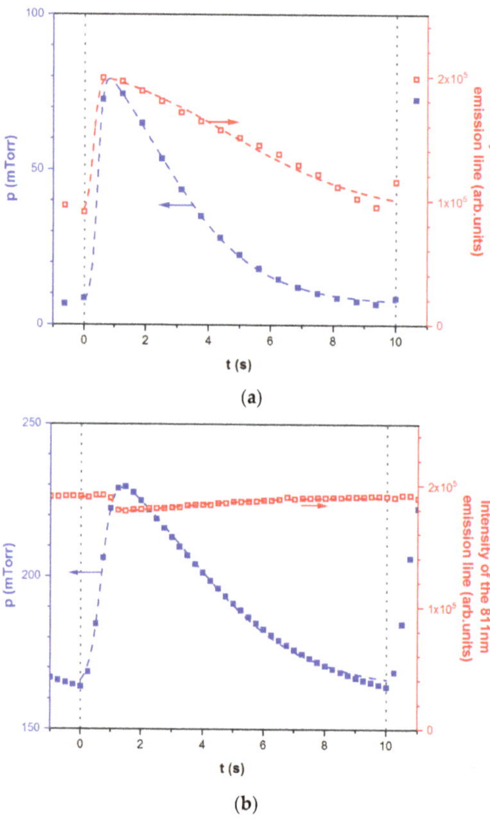

Figure 3. Comparison between the evolutions of the pressure (blue curve) and the Ar 811 nm emission line intensity (red curve) on an injection period. Injection frequency is set to 0.1 Hz corresponding to an overpressure Δp ~70 mTorr. In (**a**), the base pressure is set at p_0 ~5 mTorr and in (**b**) at p_0 ~160 mTorr.

Clearly, for both conditions presented in Figure 3, there is a synchronized evolution of the Ar 811 nm emission line intensity with the pressure pulse. In the case of the lower pressure condition, the sudden rise and slow drop in pressure led to a similar evolution of the line intensity. However, the opposite behaviour is observed at higher pressure, although the variations appear to be less pronounced than in the first condition. To understand these apparently conflicting results, it is worth looking first at the well-known equation for the measured intensity of a given emission line I_λ:

$$I_\lambda = f(\lambda) A_{ij} n_i \theta_{ij} \tag{1}$$

where $f(\lambda)$ is the apparatus function at the wavelength λ of interest, A_{ij} is the Einstein coefficient for spontaneous emission of the transition, n_i is the number density of the Ar emitting level, and θ_{ij} is the escape factor of the transition. In optically thin media ($\theta_{ij} = 1$), given that $f(\lambda)$ is measurable and that A_{ij} is a known constant, the line emission intensity therefore only depends on the number density of the emitting level. In complex media such as non-equilibrium plasmas, n_i cannot simply be calculated assuming a Boltzmann equilibrium (as in equilibrium plasmas), but requires a full description of its population and depopulation mechanisms through a particle balance equation. A schematic of the Ar 2p-to-1s transitions with the dominant population and depopulation mechanisms of Ar 2p states is presented in Figure 4.

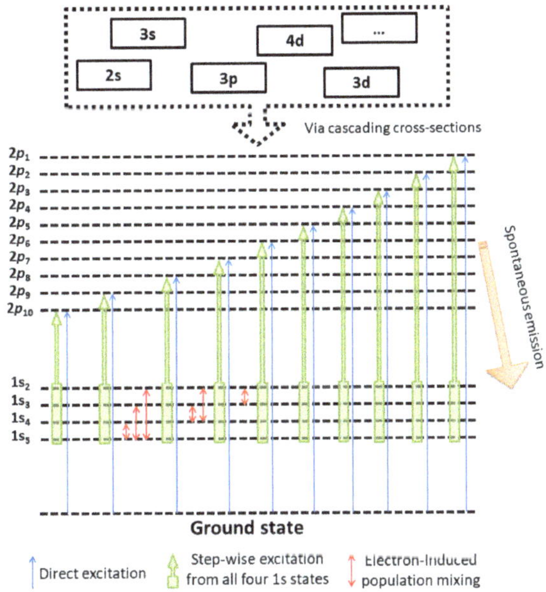

Figure 4. Schematic of the reactions considered in the CR model of argon 2p states.

As a first approximation, assuming that the Ar 2p level is mostly populated by electron impact on both ground-state argon atoms (direct excitation) and Ar 1s states (stepwise excitation via metastable and resonant levels), while only lost by spontaneous emission, the steady-state n_i can be written as:

$$n_i = \frac{k_{ground-i}\, n_{Ar} n_e + k_{step-i}\, n_{Ar^m} n_e}{\sum_j A_{ij}} \tag{2}$$

where $k_{ground\text{-}I}$ and $k_{step\text{-}I}$ are the reaction rates for direct and stepwise excitation, respectively, n_{Ar} is the number density of argon atoms in the ground state level, n_{Ar^m} is the number density of argon atoms in a metastable or resonant state, n_e is the electron number density, and $\sum_j A_{ij}$ is the sum of the Einstein coefficients corresponding to the allowed radiative transitions from the corresponding emitting level. Therefore, based on Equation (2), the opposing trends observed in Figure 3 can *a priori* be linked to changes in n_{Ar^m}, n_{Ar}, n_e, T_e or a combination of all these. This readily justifies the use of a detailed collisional-radiative model for all emitting 2p states to gain further insights into the physics driving such transient plasmas.

3.2. Results from the Collisional Radiative Model and Comparison with the Experiments

The collisional radiative model used in this work is based on the resolution of the particle balance equations of all Ar 2p levels using T_e and n_e as the only adjustable parameters. More specifically, for each (T_e, n_e) pair, it calculates a theoretical spectrum by solving the ten Ar 2p balance equations (see Figure 4) and computing the resulting intensity of relevant emission lines using Equation (1). The considered population mechanisms are direct and stepwise electron impact excitation using cascading cross-sections (in order to account for the contribution of higher-energy levels [31]), while the depopulation mechanisms are radiative transitions (mitigated by radiative trapping of the lines, when applicable) and quenching reactions by collisions with neutral argon atoms (see Figure 4). It is worth mentioning that the four Ar 1s levels balance equations are also simultaneously solved to account for stepwise excitation and radiation trapping. More details can be found in [34]. Once the theoretical emission line intensities are obtained for every (T_e, n_e) pair, they are compared to the experimental (measured) ones by calculating a percentage standard deviation and the T_e and n_e values resulting in the best fit are assumed to correspond to the real plasma parameters. In order to run, the model requires input parameters such as the operating pressure and the neutral gas temperature (to calculate the number density of argon atoms in the ground state via the ideal gas law), as well as the absorption length along the line of sight of the optical emission spectroscopy measurement to account for optically thick Ar 2p-1s transitions [18,32,33].

Figure 5a shows typical percentage standard deviation plots calculated on a set of emission bands as a function of T_e for three n_e values of interest for lowest base pressure condition, p_0 ~5 mTorr. For these experimental conditions, a minimum of the percentage standard deviation is observed for an electron temperature of 2.6 eV no matter the value of the electron number density. Here, the chosen range of electron number density (10^{14}–10^{16} m^{-3}) is based on Langmuir probe measurements done in similar experimental conditions [30]. Figure 5b confirms the optimal agreement between the theoretical and experimental spectra as found by the CR model for T_e = 2.6 eV. Similar electron temperature values were obtained by Langmuir probes for low-pressure argon RF plasmas sustained in comparable experimental conditions (pressure, reactor dimensions) [30]. Such little influence of the electron density on the optimal T_e value observed in Figure 5a implies that the CR model is mostly independent of the electron number density over the range of experimental conditions investigated. Two physical possibilities could lead to this result: either stepwise excitation processes have no contribution on the population kinetics of the Ar 2p levels, or the stepwise processes do indeed contribute but all the other population/depopulation mechanisms also have an electron number density dependency, resulting in its vanishing from the particle balance equations of Ar 1s states. To verify which of these two hypotheses applies to the present situation, a corona model based on the work of Huddlestone et al. [35] and in which stepwise processes are neglected is also calculated.

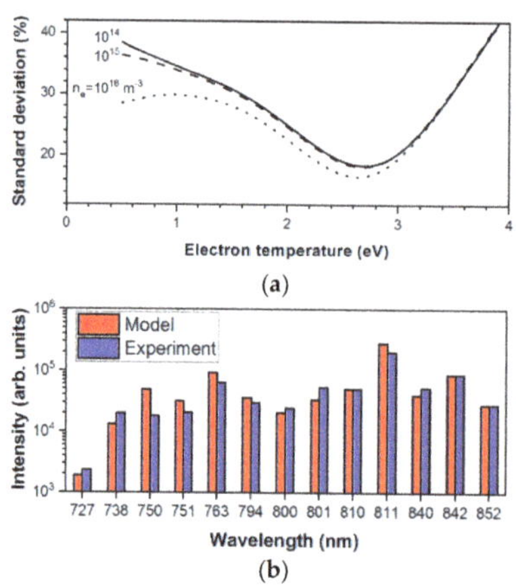

Figure 5. (**a**) Evolution of the percentage standard deviation as a function of the electron temperature for three electron number densities and a base pressure p_0 ~5 mTorr. (**b**) Agreement of the experimental and model emission line intensities for the optimal (T_e, n_e) values.

3.3. Impact of Stepwise Excitation Processes on the Study of the Electron Temperature

To determine the role of stepwise excitation processes over the range of experimental conditions examined, two separate cases were considered. First, the CR model is computed using all mechanisms that could play a role in the Ar 2p and 1s kinetics, a scenario we shall refer to as the standard CR model. Second, the same CR model is computed but all the stepwise processes are neglected, which in the end means it is considered that $n_{Ar^m} = 0$ and all the optical transition are optically thin. This scenario, based on the steady-state corona model, will be referred to as the Corona model. The temporal evolution of the electron temperature over the injection period is reported in Figure 6 for both scenarios. The evolution of the pressure drop is also reported.

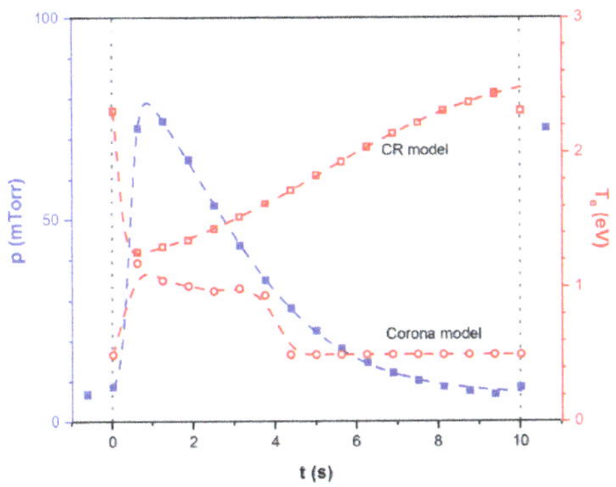

Figure 6. Comparison of the evolution of the electron temperature (red curves) and the pressure pulse (blue curve) on a normalized injection period for an injection frequency of 0.1 Hz. Electron temperatures are shown for both CR and Corona models.

In the case of the standard CR model, a decrease in T_e of about 1 eV is observed in Figure 6 concomitantly with the gas pulse, reaching a minimum of 1.4 eV and then increasing back to 2.5 eV as the pressure lowers back to its initial value. This trend agrees with the calculations of Liebermann et al. [36] as well as with the results reported by Maaloul et al. [30]. On the other hand, the opposite trend is observed in the case of the Corona model since the electron temperature rises to 1 eV and drops back to 0.5 eV as the pressure also rises and drops. More interestingly, both of the models clearly do not agree on the electron temperature at any given time of the pulse except at the close end of the pulse, where T_e (CR) is minimal and T_e (Corona) is maximal.

To verify which model better depicts the physics driving the temporal evolution of the plasma, the temporal evolution of the percentage standard deviation in both cases is analyzed; the results are shown in Figure 7a. As can be seen, the standard CR model always better describes the experimental data than the Corona model. Moreover, at the beginning of the pulse, an almost match of the electron temperature (at ~1 s) also corresponds to the time where the percentage standard deviations are equals. Therefore, even if the Corona model is less relevant to simulate the optical emission spectra, it becomes almost as accurate as the standard CR model during the gas pulse. The reason behind this similarity is due to radiation trapping, as shown in Figure 7b. Indeed, it shows the temporal evolution of the self-absorption percentage of the Ar 811 nm emission line, as calculated by the standard CR model. Self-absorption is at a minimum (~20%) directly after the beginning of the gas pulse. Since, as mentioned before, neglecting stepwise excitation processes in the Corona model implied neglecting self-absorption of argon 2p-1s transitions, it thus explains (1) why the two models almost agree when radiation trapping becomes less important, (2) why the standard CR model is systematically better at simulating the measured spectra since radiation trapping is found to be an important mechanism over an important part of the pulse cycle, and (3) why the temporal evolution of the standard deviation as well as of the electron temperature share an evolution closer to the form of the pressure pulse. Therefore, radiation trapping plays an important role in the kinetics of Ar 2p states and must imperatively be considered to obtain a correct temporal evolution of the electron temperature. Additionally, it means that processes involving Ar 1s levels do contribute to the population kinetics of the Ar 2p levels but, as mentioned before, due to all the other mechanisms also having a n_e dependency, the model is in the end independent from the electron density. Finally, a decrease in the radiation trapping with an increase in the pressure makes sense since the lower electron temperature and the additional quenching reactions by neutral argon atoms can only result in lower Ar 1s number densities and thus less self-absorption of the lines linked to 2p-to-1s transitions.

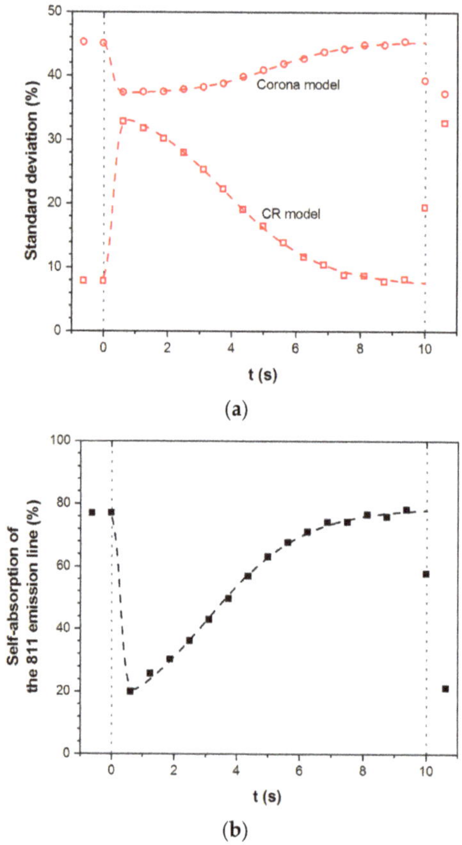

Figure 7. (**a**) Evolution of the percentage standard deviation over an injection period for the standard CR and Corona models (red curves). (**b**) Evolution of the self-absorption percentage of the Ar 811 nm emission line according to the standard CR model (black curve).

3.4. *Effect of the Pulsed Gas Injection Parameters*

Experiments with injection frequencies from 0.1 to 1 Hz are also carried out. For a fixed pumping valve position, the frequency reduces the overpressure from 65 to <1 mTorr when increasing the frequency (Figure 8). Spectra are analyzed in such conditions with the standard CR model. The temporal evolutions of the electron temperature are reported in Figure 8 for the four conditions. For an injection frequency of 0.1 Hz (p_0 ~160 mTorr), the electron temperature is, as in lower pressure condition (Figure 6), following the opposite trend of the pressure, but on a much smaller scale (Figure 8a). The experiment carried out with an injection frequency of 0.5Hz (p_0 ~110 mTorr) result in barely noticeable variation of the electron temperature with the pressure pulse (Figure 8b). Therefore, it can be concluded that the base pressure, p_0, more than the temporal variation, Δp, is determinant in fixing the electron temperature: this parameter being particularly sensible to temporal variations at low pressure. To verify this claim, measurements are recorded with an injection frequency of 1 Hz and two different base pressure of p_0 ~70 and 240 mTorr (Figure 8c,d). Clearly, the higher-pressure results in the lower electron temperature, confirming precedent conclusion in agreement with previous works at constant pressures [30].

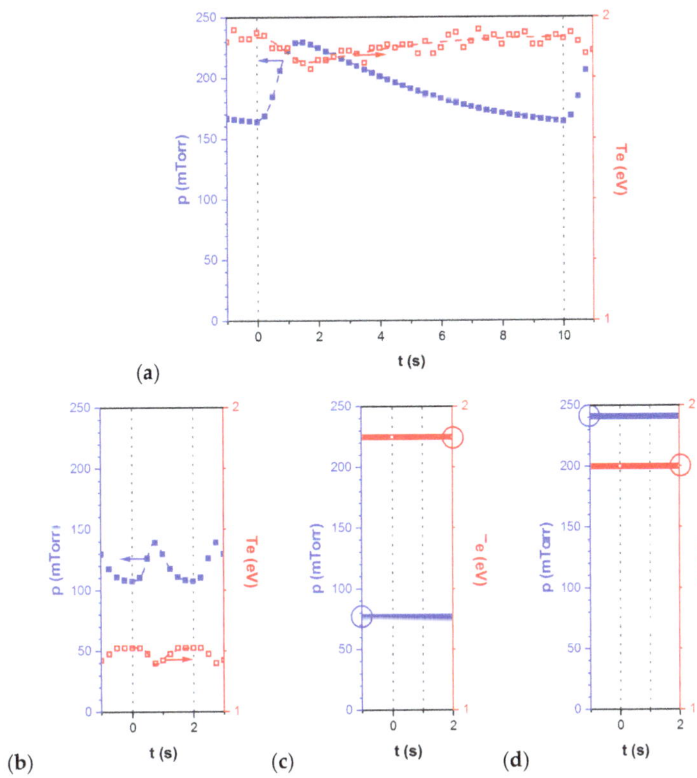

Figure 8. (**a**) Evolution of the pressure (blue curves) and of the electron temperature (red curves) during a pulse for (**a**) f = 0.1 Hz, p_0 ~160 mTorr, Δp ~70 mTorr, (**b**) f = 0.5 Hz, p_0 ~110 mTorr, Δp ~30 mTorr, (**c**) f = 1 Hz, p_0 ~70 mTorr, Δp < 1 mTorr, and (**d**) f = 1 Hz, p_0 ~240 mTorr, Δp < 1 mTorr.

3.5. Global Trend and Comparison with Reference Measurements

Figure 9 shows all the electron temperatures obtained from the comparison between measured and simulated (CR model) line emission intensities of Ar 2p-to-1s transitions under the different experimental conditions and reports them as a function of pressure. Clearly, a global trend emerges from this graph, which can be compared to a scaling law (curve). This scaling law is based on the resolution of the particle balance equation of charged species in which electrons and ions are mostly produced by electron-impact ionization of ground state argon atoms and lost by ambipolar diffusion and recombination on plasma reactor walls. In such model, the electron temperature becomes solely governed by the number density of ground state argon atoms (linked to the pressure via the ideal gas law) and the reactor dimensions. Here, the particle balance equation is solved for a cylindrical geometry with a length of 5 cm and a radius of 14 cm. Similar values are obtained from the heuristic global model reported in [36]. Based on this complete set of data, it is now clear that small variations of the pressure on the low-pressure side will affect more significantly the electron temperature than small variations of the pressure on the high-pressure side.

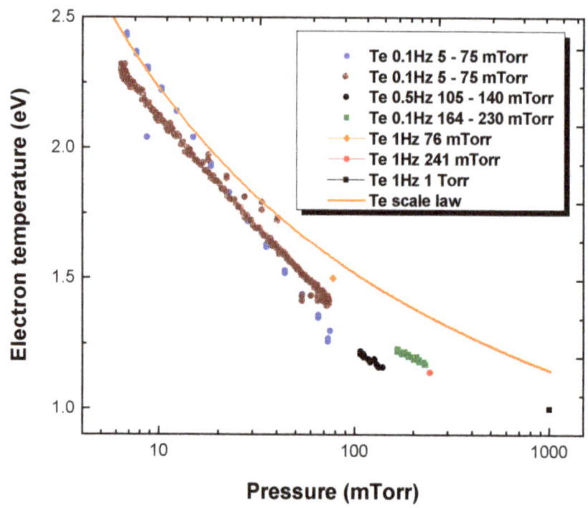

Figure 9. Evolution of the electron temperature with the pressure for the complete set of experimental conditions compared with a scale law (orange curve).

4. Conclusions and Perspectives

Optical emission spectroscopy measurement of the Ar 2p-to-1s transitions used in combination with a collisional-radiative model of the Ar 2p levels evidence that pulsed gas injections at low frequency (0.1–1 Hz) in a low-pressure capacitively coupled plasma reactor significantly affect the discharge. A decrease in the electron temperature following the pressure pulse is observed. A comparison of this result to that of a corona model highlight the importance of considering mechanisms involving Ar 1s levels, and especially radiation trapping, in the particle balance equation of Ar 2p and 1s states. Additionally, a global trend comparable to a scaling law is evidenced, linking an increase in the real-time operating pressure to a decrease in the electron temperature. Pressure variations on the low-pressure side have been observed to more importantly influence the electron temperature. This work is a first and necessary step to describe more complex plasmas in the presence of pulses of gaseous or liquid precursors with or without nanoparticles for deposition of nanocomposite coatings using the DLRI. Indeed, in such case, the role of precursors (liquid or gases) and/or nanoparticles can highly affect the plasma behavior and, especially, the electron temperature [18].

Author Contributions: Conceptualization, M.L.K., R.C. and L.S.; methodology, T.S., P.V. and G.C.; software, A.D.-J.; validation, T.S., P.V., G.C. and A.D.-J.; formal analysis, T.S.; data curation, T.S.; writing—original draft preparation, T.S., P.V. and A.D.-J.; writing—review and editing, M.L.K., R.C. and L.S.; supervision, R.C. and L.S.; project administration, R.C. and L.S.; funding acquisition, M.L.K., R.C. and L.S. All authors have read and agreed to the published version of the manuscript.

Funding: This work was financially supported by the National Science and Engineering Research Council and the Canada Research Chair program. The contributions of the Université de Montréal and CNRS through their contributions to the Québec-France International Research Network on Nanomatériaux Multifonctionnels Contrôlés (IRN-NMC) is also acknowledged.

Data Availability Statement: The data that support the findings of this study are available upon request from the authors.

Conflicts of Interest: The authors declare no conflict of interest.

References

1. Zheng, X.; Li, J.; Hu, J.; Li, J.; Ding, R.; Cao, B.; Wu, J. Comparison between gas puffing and supersonic molecular beam injection in plasma density feedback experiments in EAST. *Plasma Phys. Control. Fusion* **2013**, *55*, 115010. [CrossRef]
2. Cacot, L.; Carnide, G.; Kahn, M.L.; Caquineau, H.; Clegereaux, R.; Naude, N.; Stafford, L. Influence of pulsed gas injections on the stability of Townsend dielectric barrier discharges in nitrogen at atmospheric pressure. *J. Phys. D Appl. Phys.* **2022**, *55*, 445204. [CrossRef]
3. Ziemer, J.; Cubbin, E.; Choueiri, E.; Birx, D. Performance characterization of a high efficiency gas-fed pulsed plasma thruster. In Proceedings of the 33rd Joint Propulsion Conference and Exhibit, Seattle, WA, USA, 6–9 July 1997; p. 2925.
4. Koizumi, H.; Kakami, A.; Furuta, Y.; Komurasaki, K.; Arakawa, Y. Liquid propellant pulsed plasma thruster. In Proceedings of the 28th International Electric Propulsion Conference, Toulouse, France, 17–21 March 2003; pp. 2003–2087.
5. Kakami, A.; Koizumi, H.; Komurasaki, K.; Arakawa, Y. Design and performance of liquid propellant pulsed plasma thruster. *Vacuum* **2004**, *73*, 419–425. [CrossRef]
6. Vahlas, C.; Guillon, H.; Senocq, F.; Caussat, B.; Bonnafous, S. Solvent free method for intense vaporization of solid molecular and inorganic compounds. *Gases Instrum.* **2009**, 8–11.
7. Cacot, L.; Carnide, G.; Kahn, M.L.; Clergereaux, R.; Naudé, N.; Stafford, L. Kinetics driving thin-film deposition in dielectric barrier discharges using a direct liquid injector operated in a pulsed regime. *J. Phys. D Appl. Phys.* **2022**, *55*, 475202. [CrossRef]
8. Mitronika, M.; Profili, J.; Goullet, A.; Gautier, N.; Stephant, N.; Stafford, L.; Granier, A.; Richard-Plouet, M. TiO_2–SiO_2 nanocomposite thin films deposited by direct liquid injection of colloidal solution in an O_2/HMDSO low-pressure plasma. *J. Phys. D Appl. Phys.* **2020**, *54*, 085206. [CrossRef]
9. Pégourié, B.; Tsitrone, E.; Dejarnac, R.; Bucalossi, J.; Martin, G.; Gunn, J.; Frigione, D.; Reiter, D.; Ghendrih, P.; Clément, C. Supersonic gas injection on Tore Supra. *J. Nucl. Mater.* **2003**, *313*, 539–542. [CrossRef]
10. Pánek, R.; Gunn, J.; Bucalossi, J.; Duran, I.; Geraud, A.; Hron, M.; Loarer, T.; Pégourié, B.; Stöckel, J.; Tsitrone, E. The response of the Tore Supra edge plasma to supersonic pulsed gas injection. *J. Nucl. Mater.* **2005**, *337*, 530–534. [CrossRef]
11. Wicher, B.; Chodun, R.; Trzcinski, M.; Langier, K.N.; Skowroński, Ł.; Lachowski, A.; Zdunek, K. Applications insight into the plasmochemical state and optical properties of amorphous CNx films deposited by gas injection magnetron sputtering method. *Appl. Surf. Sci.* **2021**, *565*, 150540. [CrossRef]
12. Zdunek, K.; Nowakowska-Langier, K.; Chodun, R.; Kupczyk, M.; Siwak, P. Properties of TiN coatings deposited by the modified IPD method. *Vacuum* **2010**, *85*, 514–517. [CrossRef]
13. Subramonium, P.; Kushner, M.J. Pulsed plasmas as a method to improve uniformity during materials processing. *J. Appl. Phys.* **2004**, *96*, 82–93. [CrossRef]
14. Voitenko, D.; Ananyev, S.S.; Astapenko, G.I.; Basilaia, A.D.; Markolia, A.I.; Mitrofanov, K.N.; Myalton, V.V.; Timoshenko, A.P.; Kharrasov, A.M.; Krauz, V.I. Study of plasma flows generated in plasma focus discharge in different regimes of working gas filling. *Plasma Phys. Rep.* **2017**, *43*, 1132–1146. [CrossRef]
15. Donnelly, V.; Guha, J.; Stafford, L. Critical review: Plasma-surface reactions and the spinning wall method. *J. Vac. Sci. Technol. A Vac. Surf. Film.* **2011**, *29*, 10801. [CrossRef]
16. Nagatsu, M.; Kitagawa, N.; Takada, N. Production of Vanadium–Carbon Clusters Using DC Arc Discharges with Pulsed Helium/Methane Gas Injection. *Jpn. J. Appl. Phys.* **2000**, *39*, 6037. [CrossRef]
17. Despax, B.; Raynaud, P. Deposition of 'polysiloxane' thin films containing silver particles by an RF asymmetrical discharge. *Plasma Process. Polym.* **2007**, *4*, 127–134. [CrossRef]
18. Garofano, V.; Berard, R.; Boivin, S.; Joblin, C.; Makasheva, K.; Stafford, L. Multi-scale investigation in the frequency domain of Ar/HMDSO dusty plasma with pulsed injection of HMDSO. *Plasma Sources Sci. Technol.* **2019**, *28*, 55019. [CrossRef]
19. Kushner, M.J. Pulsed plasma-pulsed injection sources for remote plasma activated chemical vapor deposition. *J. Appl. Phys.* **1993**, *73*, 4098–4100. [CrossRef]
20. Durand, C.; Vallée, C.; Loup, V.; Salicio, O.; Dubourdieu, C.; Blonkowski, S.; Bonvalot, M.; Holliger, P.; Joubert, O. Metal–insulator–metal capacitors using Y_2O_3 dielectric grown by pulsed-injection plasma enhanced metalorganic chemical vapor deposition. *J. Vac. Sci. Technol. A Vac. Surf. Film.* **2004**, *22*, 655–660. [CrossRef]
21. Kamigaito, O. What can be improved by nanometer composites? *J. Jpn. Soc. Powder Powder Metall.* **1991**, *38*, 315–321. [CrossRef]
22. Kelly, A. *Concise Encyclopedia of Composite Materials*; Elsevier: Amsterdam, The Netherlands, 2012.
23. Mitronika, M.; Granier, A.; Goullet, A.; Richard-Plouet, M. Hybrid approaches coupling sol–gel and plasma for the deposition of oxide-based nanocomposite thin films: A review. *SN Appl. Sci.* **2021**, *3*, 665. [CrossRef]
24. Kahn, M.; Champouret, Y.; Clergereaux, R.; Vahlas, C.; Mingotaud, A.; Carnide, G. Process for the Preparation of Nanoparticles. WO2018019862A1, 26 July 2017.
25. Garofano, V.; Stafford, L.; Despax, B.; Clergereaux, R.; Makasheva, K. Cyclic evolution of the electron temperature and density in dusty low-pressure radio frequency plasmas with pulsed injection of hexamethyldisiloxane. *Appl. Phys. Lett.* **2015**, *107*, 183104. [CrossRef]
26. Garofano, V.R.; Stafford, L.; Despax, B.; Clergereaux, R.; Makasheva, K. Time-resolved optical emission spectroscopy of dusty low-pressure RF plasmas with pulsed injection of hexamethyldisiloxane. In Proceedings of the 22nd International Symposium on Plasma Chemistry, Antwerp, Belgium, 5–10 July 2015; p. 3.

27. Schweigert, I.; Alexandrov, A.L.; Ariskin, D.A.; Peeters, F.M.; Stefanović, I.; Kovačević, E.; Berndt, J.; Winter, J. Effect of transport of growing nanoparticles on capacitively coupled rf discharge dynamics. *Phys. Rev. E* **2008**, *78*, 26410. [CrossRef]
28. Chang, J.; Coburn, J. Plasma-surface interactions. *J. Vac. Sci. Technol. A Vac. Surf. Film.* **2003**, *21*, S145–S151. [CrossRef]
29. Hegemann, D.; Hossain, M.M.; Körner, E.; Balazs, D.J. Macroscopic description of plasma polymerization. *Plasma Process. Polym.* **2007**, *4*, 229–238. [CrossRef]
30. Maaloul, L.; Gangwar, R.; Morel, S.; Stafford, L. Spatially resolved electron density and electron energy distribution function in Ar magnetron plasmas used for sputter-deposition of ZnO-based thin films. *J. Vac. Sci. Technol. A Vac. Surf. Film.* **2015**, *33*, 61310. [CrossRef]
31. Donnelly, V.M. Plasma electron temperatures and electron energy distributions measured by trace rare gases optical emission spectroscopy. *J. Phys. D Appl. Phys.* **2004**, *37*, R217. [CrossRef]
32. Boivin, S.; Glad, X.; Boeuf, J.; Stafford, L. Analysis of the high-energy electron population in surface-wave plasma columns in presence of collisionless resonant absorption. *Plasma Sources Sci. Technol.* **2018**, *27*, 95011. [CrossRef]
33. Boivin, S.; Glad, X.; Latrasse, L.; Sarkissian, A.; Stafford, L. Probing suprathermal electrons by trace rare gases optical emission spectroscopy in low pressure dipolar microwave plasmas excited at the electron cyclotron resonance. *Phys. Plasmas* **2018**, *25*, 93511. [CrossRef]
34. Malyshev, M.; Donnelly, V. Trace rare gases optical emission spectroscopy: Nonintrusive method for measuring electron temperatures in low-pressure, low-temperature plasmas. *Phys. Rev. E* **1999**, *60*, 6016. [CrossRef]
35. Huddlestone, R.H.; Leonard, S.L. *Plasma Diagnostic Techniques*; Academic Press: Cambridge, MA, USA, 1965; Volume 65.
36. Lee, C.; Lieberman, M. Global model of Ar, O_2, Cl_2, and Ar/O_2 high-density plasma discharges. *J. Vac. Sci. Technol. A Vac. Surf. Film.* **1995**, *13*, 368–380. [CrossRef]

atoms MDPI

Article

Diagnostics of Argon Plasma Using Reliable Electron-Impact Excitation Cross Sections of Ar and Ar^+

Neelam Shukla [1,*,†], Reetesh Kumar Gangwar [2] and Rajesh Srivastava [1,†]

1 Department of Physics, Indian Institute of Technology Roorkee, Roorkee 247667, India
2 Department of Physics & CAMOST, Indian Institute of Technology Tirupati, Tirupati 517619, India
* Correspondence: shuklan@unk.edu
† These authors contributed equally to this work.

Abstract: Comprehensive collisional radiative (CR) models have been developed for the diagnostic of argon plasma using Ar and Ar^+ emission lines. The present CR models consist of 42 and 114 fine-structure levels of Ar and Ar^+, respectively. Various populating and depopulating mechanisms are incorporated in the model. A complete set of electron-impact fine-structure resolved excitation cross-sections for different excited levels in Ar and Ar^+ are used, which are obtained by employing relativistic distorted wave theory. Along with this, the electron-impact ionization, radiation trapping, diffusion, and three-body recombination are also considered. Further, to demonstrate the applicability of the present CR model, we applied it to characterize the Helicon-plasma utilizing the optical emission spectroscopy measurements. The key plasma parameters, such as electron density and electron temperature, are obtained using their measured Ar and Ar^+ emission line intensities. Our results are in reasonable agreement with their anticipated estimates. The matching of our calculated intensities of the different Ar and Ar^+ lines shows excellent agreement with the measured intensities at various powers.

Keywords: Ar and Ar^+ collisional radiative model; relativistic distorted wave theory; electron-impact excitation cross-sections; argon plasma

Citation: Shukla, N.; Gangwar, R.K.; Srivastava, R. Diagnostics of Argon Plasma Using Reliable Electron-Impact Excitation Cross Sections of Ar and Ar^+. *Atoms* **2022**, *10*, 118. https://doi.org/10.3390/atoms10040118

Academic Editor: Jean-Christophe Pain

Received: 26 August 2022
Accepted: 20 October 2022
Published: 24 October 2022

1. Introduction

Non-invasive spectroscopic plasma diagnostic utilizing the optical emission spectroscopy (OES) measurements offers a powerful tool to study the deriving physics of plasma kinetics and gain knowledge of the production rate of different species present in the plasma. Since most of the laboratory plasmas significantly deviate from equilibrium conditions, giving rise to the need for the development of collisional radiative (CR) models to extract the information of key plasma parameters such as electron temperature (T_e) and electron density (n_e). We observe from the literature [1–3] that either the plasma or inert gases are often considered or added in small amounts in any plasma to characterize it. We find primarily that neutral argon gas is commonly used, and the intensities of its emission lines are measured and utilized to characterize various plasmas. In many high-density argon plasmas, along with the neutral emission lines, significant emissions from higher ionic states of argon, particularly from its first ionic state (i.e., Ar^+), are also observed [4–7]. However, most of the CR models have used for the characterization of the plasma emission lines of the neutral argon, and not the lines of Ar^+. Further, in such types of plasma, the electron-impact processes are the dominant processes [4]. The most important are the electron-impact excitations/de-excitations from the ground and the first excited manifold levels. It is due to the fact that the population densities of the ground and the first excited manifold are much higher as compared to other higher levels. Therefore, it would be interesting and important to carry out plasma diagnostics by incorporating reliable electron-impact excitation cross sections of the various fine-structure levels of the Ar and Ar^+, along with other contributing processes in the CR models. To the best of our knowledge, it would be the first study where we presented the calculation of the plasma parameter results obtained from CR models using both the atomic and ionic emission lines.

Further, plasma with low discharge pressure, high electron density, and low ion and electron temperatures are very useful and important for practical plasma sources [8]. From this point of view, Helicon Plasma (HP) sources, which use Helicon waves (HW) [8,9], are quite promising as these sources can produce plasmas with high $n_e \approx 10^{13}$ cm^{-3} and ionization efficiency [10,11] for input power smaller than a kW at radio frequency (rf) with some magneto-static field. Thus, HP sources have various applications, such as in fusion research [12,13], material processing [14–16], plasma propulsions and thrusters [17–19], plasma Wakefield accelerators [20], etc. In the world's largest TOKAMAK fusion device, which is the International Thermonuclear Experimental Reactor (ITER), the neutral beam injection (NBI) ion sources are all inductively coupled plasma (ICP) type [21]. In ITER, ICP is generated inside the driver at rf 1 MHz and high powers up to 90 kW. However, it is beneficial to lower the value of the required rf power while maintaining the same plasma densities to obtain a more operational margin with accuracy and cost diminution. In this respect, various experimental studies [4,5,22] have shown that the Helicon heating is a very effective rf coupling technique to achieve plasma densities in Helicon discharges up to one order higher than the ICP discharges at the same rf power. With these developments, researchers worldwide have started investigating the HP source for the ITER nuclear beam system [11,21–26]. In most HP source [4] experiments, the working noble gas is argon; thus, it makes the diagnostics of argon plasma in these sources important and worth exploring.

In this regard, Soltani and Habibi [4] recently designed and developed an HP source for the NBI system. In their work, they identified and measured the intensities of the emission lines from both the Ar and Ar^+. Thus, it would be beneficial from the application point of view to take up the diagnostic study of the HP of argon gas in light of their OES measurements [4]. Therefore, in this study, we aim to develop the detailed CR models of Ar and Ar^+ for HP in light of the experimental study of Soltani and Habibi [4]. The present CR models of Ar and Ar^+ have been developed by incorporating important atomic processes, such as electron-impact excitation/de-excitation, electron-impact ionization, radiation trapping, diffusion, and three-body recombination. It is worth mentioning that the electron-impact processes play a dominant role in plasma kinetics. Therefore, the cross-sections of these processes should be incorporated in a consistent manner [27]. However, the measurements and benchmark non-perturbative calculations are preferred, but these are available for selected transitions and also at a very limited energy range [28–33]. Mostly these considered transitions are unresolved fine-structured and thus are not suitable for developing a plasma model. However, in the plasma model, cross-sections for a large number of fine-structured transitions are required in the wide range of impact energy. Therefore, calculations from a reliable variant of the perturbative approach are often useful. In this regard, our group has reported electron-impact cross-sections for various atoms and ions using the relativistic distorted wave approach (RDW). In this RDW theory, the bound state atomic wave functions of Ar and Ar^+ are at the Dirac–Fock level obtained from GRASP2K code [34]. Further, the projectile electron wave functions are computed numerically by solving the Dirac equations, which naturally incorporates relativistic effects such as spin-orbit interaction and jj coupling. Our calculated RDW cross-sections [35–38] are in reasonable agreement with the available respective measurements [29,30,39] and non-perturbative *R-matrix* calculations [31,32], though later better B-spline R-matrix calculations [33] for excitation from the ground state to some upper excited states were also reported, which reasonably compare with our RDW cross-sections. Additionally, our reported fine-structure resolved cross-sections have been successfully used in various plasma diagnostic studies [40–42]. All the incorporated electron-impact excitation cross-sections [35,37,38,43] of Ar and Ar^+ in the present CR model are calculated from the RDW theory.

Further, we use a spectroscopic diagnostic approach which is based on comparing the peak intensity of emission lines obtained experimentally with the intensity obtained using our CR model. It would be worth mentioning here that in the present work, we have not included the line-broadening in the CR model. We have obtained the plasma parameters by employing the OES measurements of Soltani and Habibi [4] for Ar and Ar^+ emission lines. Moreover, we would like to add that Evdokimov et al. [7] have reported their OES measurements of the magnetron discharge plasma considering Ar and Ar^+ lines and presented their modeling results. They primarily developed a CR model to characterize their magnetron plasma, following our earlier reported CR model of Gangwar et al. [36] for neutral Ar plasma, and also took only a single emission line of Ar^+ (i.e., 488 nm). In their approach, they have considered a single electron-impact transition in Ar^+, which can only be suitable for magnetron plasma. However, in the case of different plasmas, where all other possible electron excitation processes that can occur from the ground and different excited states of Ar^+ and various other population transfer mechanisms are essential, this approach might not be appropriate. All these processes (mentioned above) are considered in the present CR models, and details are given in the next section.

2. Collisional Radiative Model

The particle balance equation of the present CR model, which is used to obtain the population density (n_j) as a function of electron density and electron temperature corresponding to an excited level j of any considered system, can be written as:

$$\sum_{\substack{i=1 \\ i\neq j}}^{x} k_{ij}(T_e)n_in_e + \sum_{i>j} A_{ij}^{eff} n_i + n_en_+n_ek_{+j}(T_e) - \sum_{\substack{i=1 \\ i\neq j}}^{x} k_{ji}(T_e)n_jn_e - \sum_{i<j} A_{ji}^{eff} n_j - n_jn_ek_{j+}(T_e) - n_jk_j^{diff} = 0. \quad (1)$$

All the positive and negative terms, shown in Equation (1), represent the population and depopulation channels for the excited level j, respectively. The first and fourth terms are the population transfer by electron collisional excitation and de-excitation, respectively, whereas the second and fifth terms denote the radiative decays, respectively, from the upper levels and to the lower levels. The third and sixth terms represent the three-body recombination and ionization processes, respectively. The last term of the equation shows the depopulation by diffusion of excited states through the chamber walls. In Equation (1), x refers to the number of fine-structure (FS) levels taken in the CR model for the Ar/Ar^+. The selection of these levels is made such that, in each iteration, we added most of the fine structure levels of an excited atomic state configuration until the populations of the radiating excited levels no longer change significantly by further addition of the levels. In Equation (1), n_i refers to the population density of fine-structure levels and n_+ is the ion density of the first ionic state of the concerned atomic system, i.e., neutral Ar or Ar^+. Further, as mentioned above, the most important processes are the electron-impact excitations/de-excitations from the ground and the first excited manifold levels in the considered plasma conditions [4]. On the other hand, out of the two-body and three-body recombination processes, we have included only the latter, as it is crucial for low-temperature plasma [44]. The charge exchange process is also not included in the present model, though this can affect the charge-state balance of the low-temperature and low-density plasma. However, we omit these processes for the sake of simplicity, noting that the purpose of this study is to demonstrate the utility and accuracy of our recently calculated cross-sections. In fact, the laser-produced plasma in [45–47] is at a lower temperature (T_e~1 eV) than the present Ar plasma studies and significantly higher electron density (n_e~1E16–1E17 cm^{-3}, therefore enhancing three-body recombination), yet they still found the significance of radiative recombination. Further, the population transfer due to electron-impact excitation and ionization in Equation (1), are included through rate-coefficients k_{ij} and k_{j+}, respectively.

The rate-coefficients for electron-impact excitation can be evaluated using the following expression:

$$k_{ij} = \sqrt{\frac{2}{m_e}} \int_{E_{ij}}^{\infty} \sigma_{ij}(E)\sqrt{E} f(E) dE. \tag{2}$$

Here, m_e is the mass of electrons, E_{ij} denotes the excitation threshold energy of $i \rightarrow j$ transition. $\sigma_{ij}(E)$ is the electron-impact excitation cross-section for the $i \rightarrow j$ transition. $f(E)$ represents the Maxwellian electron energy distribution function (EEDF), which is related to the electron energy probability function through $f(E) = E^{1/2}F(E)$. This choice of EEDF is made for the sake of simplicity and to be in consistency with the experimental plasma of Soltani and Habibi [4], where they have also taken the Maxwellian EEDF. Similarly, the rate-coefficient for the electron-impact ionization can be obtained by replacing the upper-state with ionization level in Equation (2) and using electron-impact ionization cross-section (σ_{i+}) in place of excitation cross-section. The rate-coefficients for reverse processes, e.g., electron-impact de-excitation and three-body recombination, are incorporated using the detailed balance principle [41,44,48]. Thus, the electron-impact de-excitation rate-coefficient (k_{ji}) can be written in terms of excitation cross-section as:

$$k_{ji} = \frac{g_i}{g_j}\sqrt{\frac{2}{m_e}} \int_{E_{ij}}^{\infty} \sigma_{ij}(E) E F(E - E_{ij}) dE. \tag{3}$$

Here, g_i and g_j are the statistical weights of the initial and final states, respectively. Further, the rate-coefficient for the three-body recombination (k_{+i}) can be obtained by using the Saha relation [44] as follows:

$$k_{+i} = \frac{g_i}{2g_+}\left(\frac{h^2}{2\pi m T_x}\right)^{3/2} \sqrt{\frac{2}{m_e}} \int_{E_{i+}}^{\infty} \sigma_{i+}(E) E F(E - E_{i+}) dE. \tag{4}$$

Here, g_+ refers to the statistical weight of the singly ionized state of Ar/Ar^+ ground state and E_{i+} is the ionization threshold energy of Ar/Ar^+. In order to account for the radiation trapping effect, the transition probability (A_{ij}) in Equation (1) is replaced by effective transition probability (A_{ij}^{eff}), having relation $A_{ij}^{eff} = \Lambda_{ij} \times A_{ij}$, for $i \rightarrow j$ transition. Here, the escape factor (Λ_{ij}) is a function of gas temperature, dimensions of the plasma chamber [4], and population density of the lower level. We have used Mewe's approximation [49] to calculate the escape factor in the present work as given below,

$$\Lambda(K_{ij}\rho) = \frac{2 - e^{-K_{ij}\rho/1000}}{1 + K_{ij}\rho}. \tag{5}$$

The required values of transition probabilities are taken from the NIST database [50]. The radius of the plasma chamber (ρ) is taken as 2.4 cm, and the gas temperature is equal to room temperature. We solve first the particle balance equation (Equation (1)) to obtain the population densities of different excited fine-structure levels of Ar. We have taken 42 fine-structure levels (i.e., $x = 42$). These levels are the ground state ($3p^6$) and the excited configurations $3p^54s$ (with 4 FS), $3p^54p$ (with 10 FS), $3p^53d$ (with 12 FS), $3p^55s$ (with 4 FS), $3p^55p$ (with 10 FS) states, as well as the first ionization state of neutral argon, i.e., $3p^5$. All of these states are given in Table 1 in the more familiar LS coupling notation, which we specifically derived for the purpose of the present work. Additionally, these were reported elsewhere in Paschen notation [36]. Table 1 also shows the electronic configuration of the different states and their energies [50]. All these considered levels are connected through different radiative and collisional transitions such as electron-impact excitation/de-excitation, ionization, diffusion, and three-body recombination, as shown in Figure 1. In the case of neutral argon, n_+ is the population density of Ar^+ in Equation (1). The electron-impact excitation rate-coefficients are obtained using Equation (2) by employing RDW electron-impact excitation cross-sections calculated by our group [36]. In order to evaluate the ionization rate-coefficient, the required electron-impact ionization cross-sections are taken from the experiments [51,52]. The rate-coefficients for the diffusion of metastable states of Ar are taken from Kolts et al. [53]. It is worth mentioning that the escape factors required in Equation (1) are computed for the transitions which decay to the ground state of Ar, as only these have a significant contribution compared to other transitions.

Table 1. The energy levels of Ar, considered in the present CR model.

Level No.	Level	LS Coupling	Excitation Energy (eV)
1	$3p^6$	$^1S_0$	0
2	$3p^5(^2P^{\circ}_{3/2})4s$	$^3P_2$	11.548
3		$^3P_1$	11.623
4	$3p^5(^2P^{\circ}_{1/2})4s$	$^3P_0$	11.723
5		$^1P_1$	11.828
6		$^3D_1$	12.907
7		$^3D_3$	13.076
8	$3p^5(^2P^{\circ}_{3/2})4p$	$^3D_2$	13.095
9		$^3P_1$	13.153
10		$^3P_2$	13.172
11		$^3P_0$	13.273
12		$^3S_1$	13.283
13	$3p^5(^2P^{\circ}_{1/2})4p$	$^1D_2$	13.302
14		$^1P_1$	13.328
15		$^1S_0$	13.480
16		$^3P_0$	13.845
17		$^3P_1$	13.863
18		$^3P_2$	13.903
19	$3p^5(^2P^{\circ}_{3/2})3d$	$^3F_4$	13.979
20		$^3F_3$	14.012
21		$^3F_2$	14.063
22		$^3D_3$	14.100
23		$^3D_1$	14.153
24		$^3D_2$	14.214
25	$3p^5(^2P^{\circ}_{1/2})3d$	$^1D_2$	14.234
26		$^1F_3$	14.236
27		$^1P_1$	14.303
28	$3p^5(^2P^{\circ}_{3/2})5s$	$^3P_2$	14.068
29		$^3P_1$	14.090
30	$3p^5(^2P^{\circ}_{1/2})5s$	$^3P_0$	14.241
31		$^1P_1$	14.255
32		$^3D_1$	14.464
33		$^3D_3$	14.500
34	$3p^5(^2P^{\circ}_{3/2})5p$	$^3D_2$	14.506
35		$^3P_1$	14.525
36		$^3P_2$	14.529
37		$^3P_0$	14.575
38		$^3S_1$	14.680
39	$3p^5(^2P^{\circ}_{1/2})5p$	$^1P_1$	14.687
40		$^1D_2$	14.688
41		$^1S_0$	14.738
42	$3p^5$	$^2P_{3/2}$	15.760

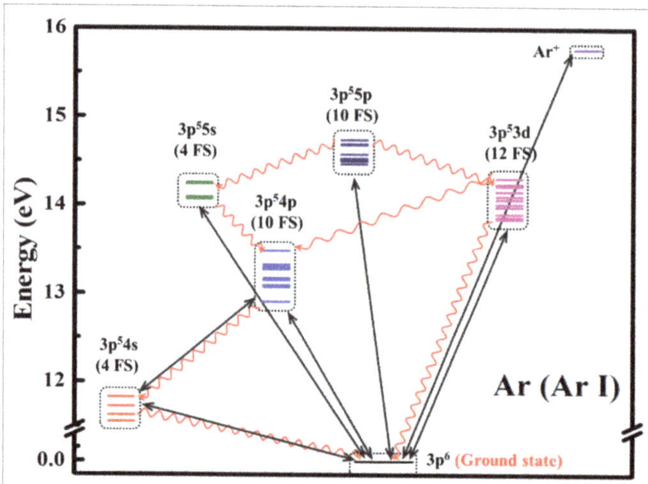

Figure 1. Energy level diagram of Ar along with the various collisional and radiative processes considered in the present CR model. The solid lines show excitations and de-excitation from the ground state of Ar. Wavy lines represent the radiative transitions. Here, FS stands for fine-structure.

In the next step, we solve the particle balance equation (Equation (1)) to obtain the population densities of different excited fine-structure levels of Ar^+. We considered 114 fine-structure levels (i.e., $x = 114$). These are $3p^5$ (ground state) and excited $3s3p^6$, $3p^43d$ (with 28 FS), $3p^44s$ (with 8 FS), $3p^44p$ (with 21 FS), $3p^45s$ (with 8 FS), $3p^44d$ (with 26 FS), $3p^45p$ (with 19 FS) states as well as first ionization level of Ar^+ $3p^4$. These considered levels are listed in Table 2, along with their electronic configurations and associated energies [50]. The CR model framework outlining the interconnection between FS levels via different radiative and collisional transitions is shown in Figure 2, such as electron-impact excitation/de-excitation, electron-impact ionization, and three-body recombination. Here, n_+ is the population density of Ar^{++} in Equation (1). The required RDW cross-sections of 114 fine-structure levels to evaluate electron-impact excitation rate-coefficient (Equation (2)) are taken from our previous work [38]. In addition, we have also calculated the additional RDW electron-impact excitation cross-sections of $3p^44s \rightarrow 3p^44p$ transitions because we did not report these cross-sections earlier [38]. These transitions play a vital role in the population and depopulation of the considered emission lines in the present work. The electron-impact ionization cross-sections of Ar^+ needed to calculate the ionization rate-coefficients are taken from the reported measurements [54]. The required radiation trappings in Equation (1) are calculated for the transitions which decay to the ground state of Ar^+, as we have found that only for these transitions, giving significant contributions.

Table 2. The energy levels of Ar^+, considered in present CR model.

Level No.	Level	LS Coupling	Excitation Energy (eV)	Level No.	Level	LS Coupling	Excitation Energy (eV)
1	$3p^5$	$^2P_{3/2}$	0	59	$3p^4(^1S)4p$	$^2P_{3/2}$	23.802
2		$^2P_{1/2}$	0.177	60		$^2P_{1/2}$	23.846

Table 2. *Cont.*

Level No.	Level	LS Coupling	Excitation Energy (eV)	Level No.	Level	LS Coupling	Excitation Energy (eV)
3	**$3s3p^6$**	$^2S_{1/2}$	13.480	61		$^4P_{5/2}$	22.515
4		$^4D_{7/2}$	16.406	62		$^4P_{3/2}$	22.592
5		$^4D_{5/2}$	16.425	63	**$3p^4(^3P)5s$**	$^4P_{1/2}$	22.683
6		$^4D_{3/2}$	16.444	64		$^2P_{3/2}$	22.700
7		$^4D_{1/2}$	16.457	65		$^2P_{1/2}$	22.802
8		$^4F_{9/2}$	17.629	66	**$3p^4(^1D)5s$**	$^2D_{5/2}$	24.284
9		$^4F_{7/2}$	17.695	67		$^2D_{3/2}$	24.284
10		$^4F_{5/2}$	17.743	68	**$3p^4(^1S)5s$**	$^2S_{1/2}$	26.665
11		$^4F_{3/2}$	17.776	69		$^4D_{7/2}$	22.773
12	**$3p^4(^3P)3d$**	$^2P_{1/2}$	17.942	70		$^4D_{5/2}$	22.788
13		$^2P_{3/2}$	18.061	71		$^4D_{3/2}$	22.811
14		$^4P_{1/2}$	18.254	72		$^4D_{1/2}$	22.837
15		$^4P_{3/2}$	18.288	73		$^4F_{9/2}$	22.949
16		$^4P_{5/2}$	18.334	74		$^4F_{7/2}$	22.014
17		$^2F_{7/2}$	18.496	75		$^4F_{5/2}$	23.070
18		$^2F_{5/2}$	18.616	76		$^4P_{1/2}$	23.082
19		$^2D_{3/2}$	18.656	77	**$3p^4(^3P)4d$**	$^4F_{3/2}$	23.103
20		$^2D_{5/2}$	18.732	78		$^4P_{3/2}$	23.119
21		$^2G_{9/2}$	19.116	79		$^2F_{7/2}$	23.162
22		$^2G_{7/2}$	19.119	80		$^4P_{5/2}$	23.171
23		$^2F_{5/2}$	20.246	81		$^2F_{5/2}$	23.258
24	**$3p^4(^1D)3d$**	$^2F_{7/2}$	20.272	82		$^2P_{1/2}$	23.549
25		$^2D_{5/2}$	21.367	83		$^2P_{3/2}$	23.630
26		$^2D_{3/2}$	21.428	84		$^2D_{5/2}$	23.874
27		$^2P_{3/2}$	21.624	85		$^2D_{3/2}$	23.893
28		$^2P_{1/2}$	21.675	86		$^2G_{7/2}$	24.623
29	**$3p^4(^1S)3d$**	$^2D_{5/2}$	22.266	87		$^2G_{9/2}$	24.624
30		$^2D_{3/2}$	22.309	88		$^2P_{1/2}$	24.728
31	**$3p^4(^1D)3d$**	$^2S_{1/2}$	22.825	89	**$3p^4(^1D)4d$**	$^2P_{3/2}$	24.738
32		$^4P_{5/2}$	16.644	90		$^2D_{5/2}$	24.757
33		$^4P_{3/2}$	16.748	91		$^2D_{3/2}$	24.794
34	**$3p^4(^3P)4s$**	$^4P_{1/2}$	16.812	92		$^2F_{7/2}$	24.814
35		$^2P_{3/2}$	17.140	93		$^2F_{5/2}$	24.826
36		$^2P_{1/2}$	17.266	94	**$3p^4(^1D)4d$**	$^2S_{1/2}$	25.445
37	**$3p^4(^1D)4s$**	$^2D_{3/2}$	18.426	95		$^4P_{5/2}$	23.404
38		$^2D_{5/2}$	18.454	96		$^4P_{3/2}$	23.438
39	**$3p^4(^1S)4s$**	$^2S_{1/2}$	20.743	97		$^4D_{7/2}$	23.484
40		$^4P_{5/2}$	19.223	98		$^4P_{1/2}$	23.487
41		$^4P_{3/2}$	19.261	99		$^4D_{5/2}$	23.514
42		$^4P_{1/2}$	19.305	100		$^4D_{3/2}$	23.570
43		$^4D_{7/2}$	19.494	101	**$3p^4(^3P)5p$**	$^2P_{1/2}$	23.581
44		$^4D_{5/2}$	19.549	102		$^2P_{3/2}$	23.620
45		$^4D_{3/2}$	19.610	103		$^2D_{5/2}$	23.620
46	**$3p^4(^3P)4p$**	$^4D_{1/2}$	19.642	104		$^4D_{1/2}$	23.649
47		$^2D_{5/2}$	19.680	105		$^2S_{1/2}$	23.674
48		$^2D_{3/2}$	19.762	106		$^2D_{3/2}$	23.682
49		$^2P_{1/2}$	19.801	107		$^4S_{3/2}$	23.702
50		$^2P_{3/2}$	19.867	108		$^2P_{3/2}$	25.170
51		$^4S_{3/2}$	19.967	109		$^2F_{5/2}$	25.187
52		$^2S_{1/2}$	19.972	110	**$3p^4(^1D)5p$**	$^2F_{7/2}$	25.193
53		$^2F_{5/2}$	21.127	111		$^2P_{1/2}$	25.242
54		$^2F_{7/2}$	21.143	112		$^2D_{3/2}$	25.307
55	**$3p^4(^1D)4p$**	$^2P_{3/2}$	21.352	113		$^2D_{5/2}$	25.309
56		$^2P_{1/2}$	21.426	114	**$3p^4$**	$^3P_2$	27.630
57		$^2D_{3/2}$	21.492				
58		$^2D_{5/2}$	21.498				

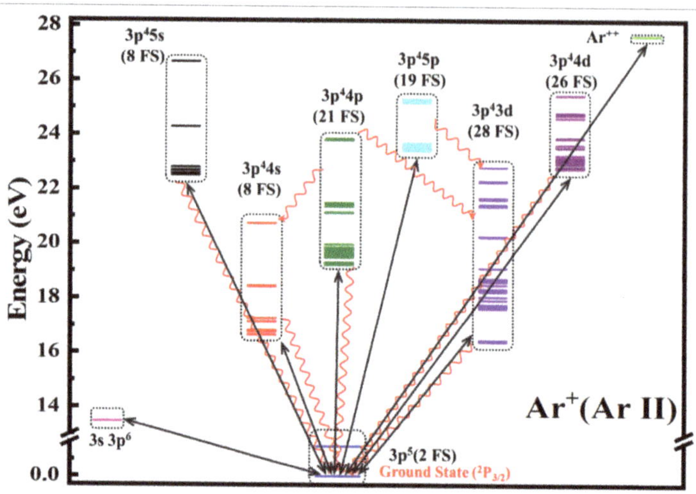

Figure 2. Energy level diagram of Ar^+ along with the various collisional and radiative processes considered in the present CR model. The solid lines show excitations and de-excitation from the ground state of Ar^+ ($^2P_{3/2}$). Wavy lines represent the radiative transitions. Here FS stands for fine-structure.

We have solved the coupled particle balance equations (Equation (1)) numerically by utilizing the standard matrix inversion approach, similar to our previous CR models [36,42]. The evaluation of the rate-coefficients of the different processes has been performed as an intermediate step by providing the necessary cross-sections and the required transition probabilities. The needed transition probabilities are taken from the NIST database [50]. Simpson's integration method is used to calculate the integrals corresponding to rate-coefficients in Equations (2)–(4). The required ground state population of the Ar atom in the CR model is evaluated from the standard gas law at the experimental conditions mentioned in Ref. [4], whereas in the case of the Ar^+, following the charge neutrality condition, we take the ground state population density as equal to the electron density that we obtained from our CR model for argon. Since Equation (1) depends on the values of T_e and n_e for solving it, we assume a tentative wide grid of the different values of T_e and n_e separately. Then, by taking each value of T_e and different values of n_e, we solve Equation (1) for each set of (T_e and n_e). After solving the coupled particle balance equations, we obtain the population densities (n_j) of each considered fine-structure level as a function of different sets of T_e and n_e in the wide range of their values. Further, using these obtained population densities, we have calculated the intensities (I) of the considered emission lines of Ar and Ar^+ using the following equation,

$$I_{ji} \propto n_j \frac{hc}{\lambda_{ji}} A_{ji}^{eff}. \tag{6}$$

Here, h and c represent Planck's constant and the speed of light, respectively. λ_{ji} ($j \rightarrow i$) stands for the emitted wavelength. Now, to obtain the plasma parameters, i.e., values of T_e and n_e which exactly correspond to the experimental plasma, the calculated CR model intensities are needed to be optimized or matched with their corresponding OES measured intensities [4]. A conventional way to achieve this optimization or matching is to calculate standard deviation parameters between the CR model simulated intensities and respective OES measured intensities.

To match our calculated model intensities for the considered emission lines of Ar and Ar^+ with the experimentally measured intensities [4], we have first normalized these intensities individually by using the following relation,

$$I^{normalized}_{j,OES(Model)} = \frac{I_{j,OES(Model)}}{\sum_{j=1}^{n'} I_{j,OES(Model)}} \times 100. \quad (7)$$

Here, $I^{normalized}_{j,OES}$ and $I^{normalized}_{j,Model}$ are the normalized intensities estimated from the OES measurements [4] and CR model, respectively. The n' represents the number of considered emission lines of Ar and Ar^+, which is four (i.e., $n' = 4$), and the choice of these lines is justified in the next section. Further, we have employed the minimum scatter or least square approximation approach to check the best match of the calculated and experimental intensities [4] by obtaining the following deviation parameter to be minimum,

$$\Delta = \sum_{j=1}^{n'} \left(I^{normalized}_{j,OES} - I^{normalized}_{j,Model}\right)^2_{T_e n_e}. \quad (8)$$

As mentioned above, the deviation parameter is a way to measure the agreement of the model intensities with the corresponding measurements [4]. Thus, its magnitude represents the least square difference between the normalized intensities obtained from the OES measurements [4] and the CR model. Therefore, the intensities obtained from the CR model for different sets of T_e and n_e are used in the above relation in order to find when the deviation parameter has the minimum value. Once this minimum condition with a particular combination of (n_e and T_e) is achieved, the corresponding value of T_e and n_e are considered to represent the actual plasma parameters. This we will refer to as the obtained/extracted parameters (T_e and n_e) for the specific case. This method has also been used successfully by us in our earlier work [36,41,42] to find the plasma parameters by matching the CR model and experimentally measured intensities.

3. Results and Discussion

Before discussing the present diagnostic results obtained from our CR models, we wish to briefly highlight the significance of using a CR Model in extracting the plasma parameters from OES measurements. The literature reveals that to extract n_e and T_e from OES measurements, the intensity of Ar-750.2 nm (for relative electron density estimation) and the ratio of Ar-811.5 and Ar-750.2 nm is often used to study the E-H-W mode transitions [55,56]. In the determination of line-ratio as a function of electron temperature, the upper radiating levels are assumed to be populated from the ground and metastable states through the electron-impact excitation process. This approach is very relevant; however, it should be noted that the line-ratio showed significant dependence only up to 3 eV [56]. Beyond that, the ratio is almost insensitive to the change in electron temperature [56]. Therefore, the scheme may not be suitable for helicon mode, where the presence of significant emission from Ar^+ suggests a relatively higher electron temperature. A recent study reported a considerable discrepancy between the electron temperature values obtained from the Langmuir probe and the line-ratio [55] in Ar Helicon plasma. Therefore, applying CR models to extract plasma parameters from OES measurements is relevant.

As mentioned above, we employed the present CR models to study the experimental argon gas plasma reported by Soltani and Habibi [4]. They designed and developed a Helicon plasma source and showed the mode change from ICP to Helicon using OES measurements. In ICP mode, the magneto-static field ($B_\circ$) is zero, whereas, in the case of Helicon mode, it varies from 350 to 750 G. Simultaneously, in both the modes, they recorded the spectra with changing rf power from 300–1000 W keeping the neutral background pressure at 0.7 mTorr. In their spectra, they have identified the specific emission lines of Ar in the range of 650–850 nm in the case of ICP mode, whereas in Helicon mode, they observed emission lines representing Ar and Ar^+ in the spectral range of 400–850 nm. In the present work, we have analyzed the data of OES measurements at $B_\circ = 0$ G (ICP mode) and $B_\circ = 750$ G (Helicon mode) of argon gas plasma provided by Soltani and Habibi [4]. It is worth mentioning that the $3p^54p$ FS levels of Ar and $3p^44p$ FS levels of Ar^+ are quite crucial for the present plasma diagnostics. The reason is that Soltani and Habibi [4] identified emission lines in their spectra of Ar and Ar^+, which corresponds to the decay from these levels. Here, we also selected the emission lines of Ar: 738.4, 751.4, 772.4, and 811.5 nm, and for Ar^+: 434.8, 458.9, 480.6, and 487.9 nm that are originating from the $3p^54p$ and $3p^44p$ FS levels of Ar and Ar^+, respectively. In addition, only these lines have substantial intensities, which are necessary for the calculation at all powers. Further, only these combinations of lines were sensitive with respect to the plasma parameters, i.e., electron temperature and electron density. Consequently, the evaluation of the population density distributions of such FS levels is the main goal of our CR models. We have used the Ar lines in the ICP mode and Ar and Ar^+ lines in Helicon mode for plasma diagnostics. The Ar emission line at 772.4 nm is unresolved consisting of two lines at 772.38 nm and 772.42 nm. Therefore, while performing comparison, the theoretical model intensity is the summed value of these two lines.

Using our CR models of Ar and Ar^+, we have calculated intensities of their emission lines as mentioned above through Equation (7) at several combinations of T_e and n_e in the wide range. These calculations are performed separately at various plasma operating powers of 300, 600, 800, 900, and 1000 W. Thereafter, the deviation parameters are calculated using the intensities of Ar lines in ICP mode and Ar as well as Ar^+ lines in Helicon mode through Equation (8). Further, these deviation parameters are plotted as a function of T_e and n_e in each mode at different powers. Then, we analyzed them to ascertain the minima to obtain the electron temperature and electron density values, which eventually characterize the plasma. As an illustration, the variation of the deviation parameter with respect to the electron temperature (at fixed n_e, which we found corresponds to the minimum deviation parameter) at 1000 W for both the modes are shown in Figure 3. From this figure, we can see that for $B_\circ = 0$ (ICP mode), the minimum value of deviation parameter is found at 4.0 eV with $n_e = 1 \times 10^{17}$ m^{-3} and for $B_\circ = 750$ G (Helicon mode) at 8.13 eV with $n_e = 1 \times 10^{19}$ m^{-3}. Thus, these values of the T_e and n_e, we can take as the extracted plasma parameters at 1000 W in different modes. It should be noted that in both cases, the minimum values of deviation parameters are less than 10 units. As explained earlier, the deviation parameter measures the agreement of the model intensities with the corresponding measurements [4]. Thus, 10 units can be said to as a good match of the intensities obtained from the CR model and experiment [4]. Similarly, we obtained the T_e and n_e values at all other powers for ICP and Helicon modes.

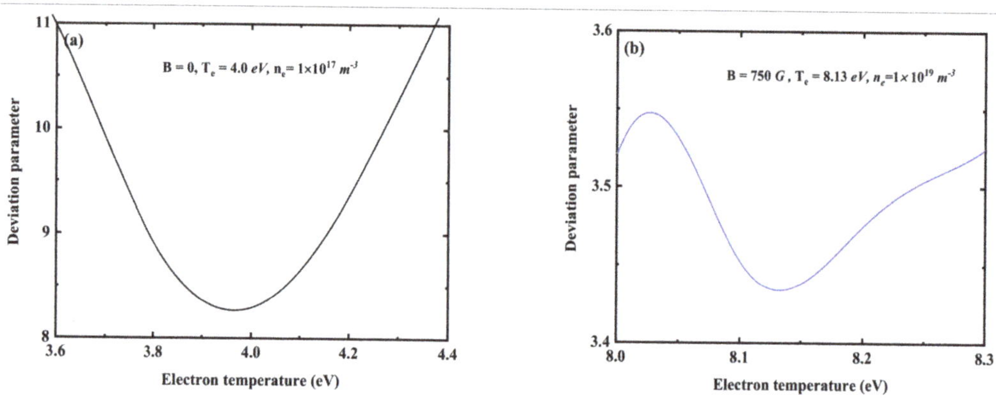

Figure 3. Deviation parameters as a function of T_e at the extracted value of n_e for argon plasma for (**a**) ICP mode (using Ar lines) and (**b**) Helicon mode (using Ar^+ lines) at 1000 W.

We have shown all the obtained plasma parameters (i.e., T_e and n_e) in both the modes in Table 3. From this table, one can observe that for the case of ICP mode, the electron temperature slightly increases from 3.39 to 4.00 eV with an increase of the power from 300 to 1000 W. Our values are quite close to the experimental value of electron temperature 4.0 eV for the ICP mode as reported by Soltani and Habibi [4]. We also found from our results that in the ICP mode, the electron density did not change much with respect to the variation in power and remained close to the value of 1×10^{17} m^{-3}. Further, our model predicts a similar feature of the variation of electron temperature with respect to the power obtained using Ar lines in Helicon mode as was in ICP. We observe that the T_e varied from 5.65 to 7.95 eV in the Helicon mode as the power increased from 300 to 1000 W. However, in contrast to ICP mode, the electron density is not constant but varies from the value of 3×10^{18} to 1×10^{19} m^{-3} with the increase of power. We also observe from the present values that when the power is increased beyond 600 W in Helicon mode (with Ar lines), the electron temperature is abruptly increased. This suggests that up to the 600 W power, the argon plasma is in ICP mode, and thereafter, the plasma gets ionized, which is in confirmation with the observation of Soltani and Habibi [4]. Because of this reason, in Helicon mode, at lower powers, i.e., 300–600 W, Ar^+ lines are not observed experimentally, whereas at higher powers, i.e., 800–1000 W range, both Ar and Ar^+ spectral lines have been observed. Since both the spectral lines from Ar and Ar^+ represent the same Helicon plasma, the plasma parameters obtained from the CR models should ideally be the same for a particular power. We also find from Table 3 that our CR models using Ar and Ar^+ lines give quite close values of T_e and n_e in the range of 800–1000 W.

Table 3. Electron temperature (T_e) and electron density (n_e) for argon plasma at different modes and powers obtained using the present CR models.

Power (W)	T_e (eV) ICP Mode (Using Ar Lines)	T_e (eV) Helicon Mode (Using Ar Lines)	T_e (eV) Helicon Mode (Using Ar^+ lines)	n_e (m^{-3}) Helicon Mode (Using Ar Lines)	n_e (m^{-3}) Helicon Mode (Using Ar^+ Lines)
300	3.39	5.56	-	3.00×10^{18}	-
600	3.59	5.95	-	4.48×10^{18}	-
800	3.61	7.39	7.96	9.79×10^{18}	9.70×10^{18}
1000	4.00 4.00 (Exp. [4])	7.95	8.13	1.00×10^{19}	1.0×10^{19}

Since Soltani and Habibi [4] have recorded the spectra with rf power variation in two different modes, it would be worth seeing how our plasma parameters behave with these variations. In Figure 4, we have shown the plot between the modes and obtained electron temperatures at 1000 W. There is a huge difference in the value of T_e obtained using Ar lines in the ICP and Helicon modes. This may be due to the fact that in Helicon mode, the plasma gets more ionized as compared to the ICP mode. The appearance of Ar^+ emission lines in Helicon mode suggests that a higher fraction of electron-neutral collisions are resulting in excited argon ions rather than excited neutrals. It is only possible if the average electron energy is higher, as the threshold excitation energy is much higher for Ar ion levels than Ar-neutral excited states. The higher average energy also supports a higher ionization degree in Helicon mode. In this mode, the electron temperatures obtained using Ar and Ar^+ lines are nearly equal as expected because the plasma is the same. Further, for the Helicon mode, the variation of electron temperatures obtained using Ar lines with respect to powers is given in Figure 5. As shown in this figure, we observe a sudden jump in the temperature, which represents the transition from ICP to Helicon mode, as pointed out earlier.

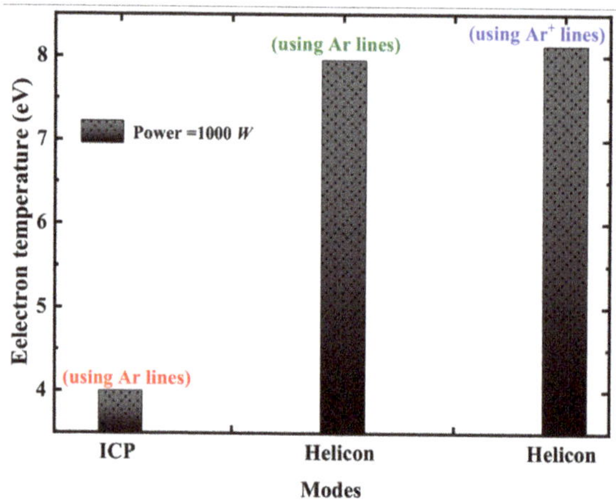

Figure 4. Variation of obtained electron temperature (T_e) using the present CR model in different plasma operating modes at fixed power of 1000 W.

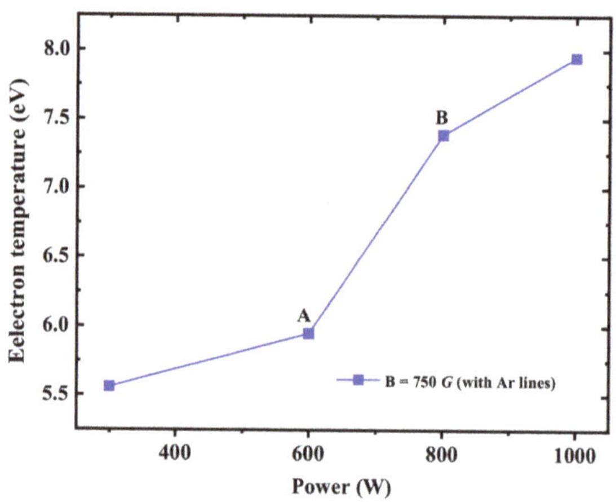

Figure 5. Variation of obtained electron temperatures (T_e) using present CR model at various powers for Helicon mode using Ar lines. The jump from point A to B reflects the transition of plasma operating mode.

Furthermore, in Figures 6–8, we have compared the normalized intensities obtained from the CR models (at the plasma parameters as given in Table 3) with the OES measurements [4]. In Figures 6 and 7, we find that the intensities of the considered transitions of Ar show good agreement with the experimental results [4] within the 4%. The largest differences are observed for the spectral lines 738.4 and 772.4 nm. Further, in Figure 8, we observe good agreement for all considered emission lines of Ar^+ with the experimental intensities [4]. The maximum difference was observed for 487.9 nm, which is 2%. Since the largest discrepancies are small in all three cases (Figures 6–8), we can say that predictions from our CR model are quite reasonable. Consequently, the obtained plasma parameters (T_e and n_e) given in Table 3 at various powers are reliable.

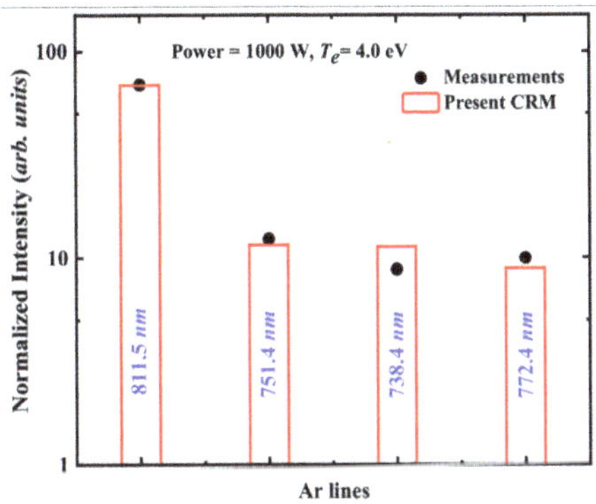

Figure 6. Comparison of normalized intensities obtained from CR model and OES measurements [4] of argon plasma in ICP mode (using Ar lines) at 1000 W.

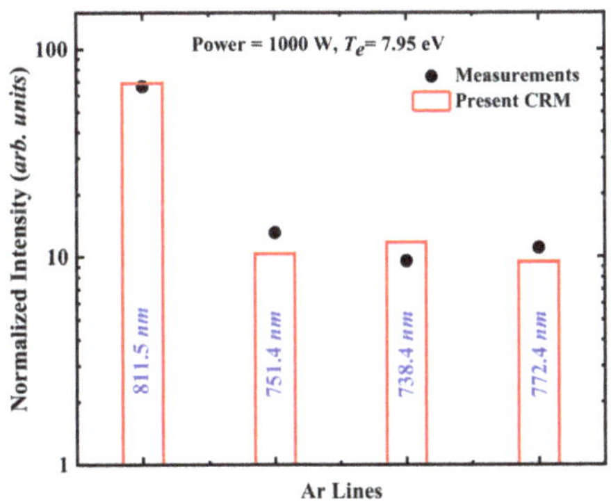

Figure 7. Comparison of normalized intensities obtained from CR model and OES measurements [4] of argon plasma in Helicon mode (using Ar lines) at 1000 W.

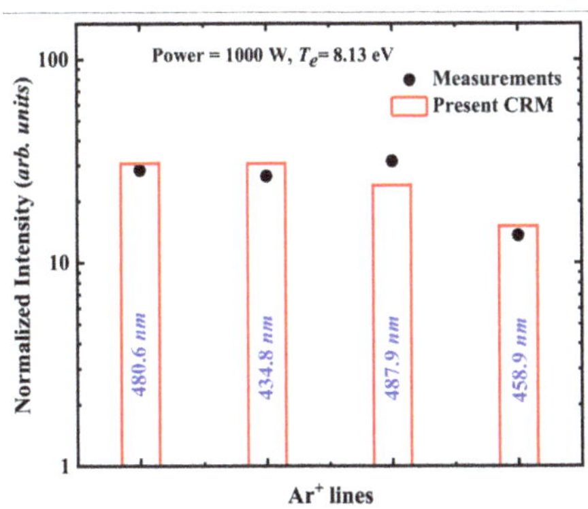

Figure 8. Comparison of normalized intensities obtained from CR model and OES measurements [4] of argon plasma in Helicon mode (using Ar^+ lines) at 1000 W.

As population densities are the main output of the CR models through which we have calculated the intensities in different modes at various powers, it would be thus worth seeing how the population densities vary with respect to the experimental values, which we have obtained through the measured intensities [4] using Equation (6). For this purpose, in Figure 9, we plotted the ratio of upper-level population densities of the considered four emission lines in the present work, with respect to the wavelengths for the ICP and Helicon modes at 1000 W. To calculate this ratio, we divided all the upper-level population densities of the considered lines of Ar (738.4, 751.4, 772.4, and 811.5 nm) and Ar^+ (434.8, 458.9, 480.6, and 487.9 nm) with the density corresponding to the largest wavelength among these lines in each case, i.e., 811.5 and 487.9 nm. The population densities with which the ratios are obtained are 3.24×10^{17} m^{-3} in ICP mode and 1.97×10^{17} m^{-3} and 1.17×10^{14} m^{-3} in Helicon mode. These are shown in Figure 9a–c, respectively. On comparing our results with the values obtained from measurements [4], we find a reasonable agreement between both sets of results for the ICP and Helicon modes at 1000 W.

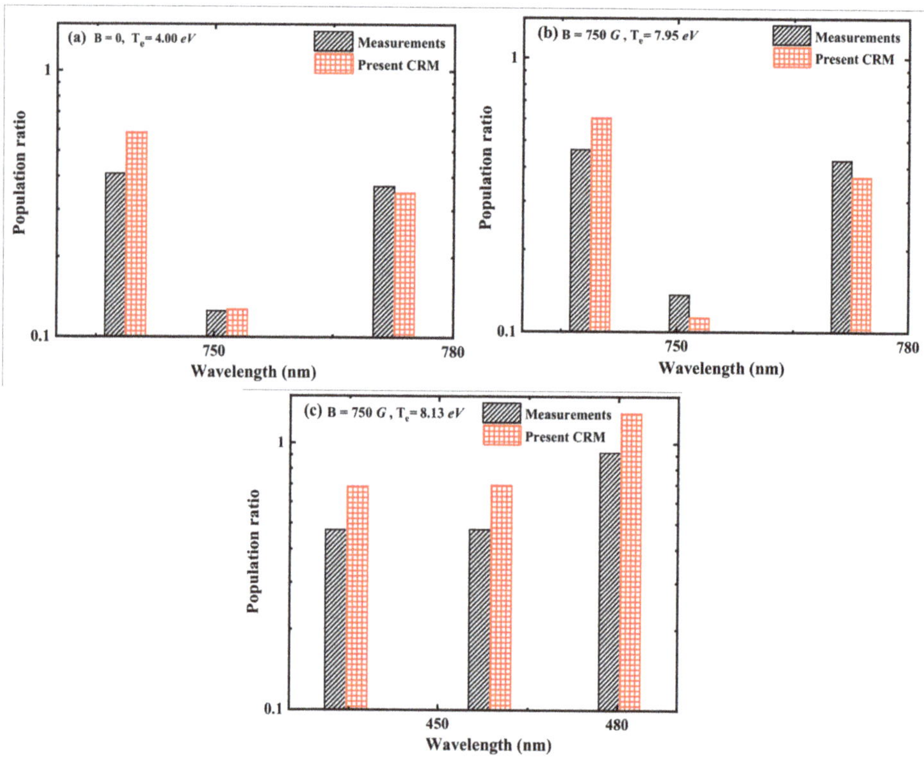

Figure 9. Comparison of upper-excited level population ratio of the considered emission lines with measurements [4] for argon plasma in (**a**) ICP mode (using Ar lines), (**b**) Helicon mode (using Ar lines), and (**c**) Helicon mode (using Ar^+ lines), at 1000 W.

4. Conclusions

In the present work, we have developed a fine-structure resolved CR model using Ar and Ar^+ emission lines for the diagnostics of the argon gas plasma. The required OES measurements are taken from the work of Soltani and Habibi [4]. Our models contain many possible populations transfer mechanisms among the fine-structure levels of Ar and Ar^+. The intensities of the different lines observed in the measurements for Ar and Ar^+ are calculated using the present CR models. These values are compared with the experimental intensities to extract the plasma parameters (T_e and n_e) in ICP and Helicon modes at various powers. The extracted plasma parameters for both these modes at different powers seem to characterize the argon gas plasma of Soltani and Habibi [4] appropriately. We have also presented and compared the upper-level population density ratios with the measured values [4] and found good agreement. We believe our present extensive CR models consider most population transfer mechanisms and can be easily extended to characterize any other Ar and Ar^+ plasma.

Author Contributions: N.S. carried out the calculations, analysed the results, and took the lead in writing the manuscript. R.K.G. was involved in the discussions. R.S. supervised the project and arranged the essential funds to perform the research. All authors contributed to finalising the manuscript. All authors have read and agreed to the published version of the manuscript.

Funding: R.S. is thankful to the SERB-DST New Delhi, Government of India for the sanction of research grant CRG/2020/005597. R.K.G. is thankful to the SERB-DST New Delhi, Government of India for financial support through grant CRG/2018/000419 and CVD/2020/000458 and acknowledges the support from CAMOST and IIT Tirupati.

Data Availability Statement: The data that support the findings of this study are available from the authors upon reasonable request.

Acknowledgments: We are grateful to Morteza Habibi for providing us their published measurements in numerical form. N.S. is thankful to the Ministry of Human Resources and Development (MHRD), Government of India for fellowship.

Conflicts of Interest: The authors declare no conflict of interest.

References

1. Malyshev, M.V.; Donnelly, V.M. Trace Rare Gases Optical Emission Spectroscopy: Nonintrusive Method for Measuring Electron Temperatures in Low-Pressure, Low-Temperature Plasmas. *Phys. Rev. E* **1999**, *60*, 6016. [CrossRef] [PubMed]
2. Donnelly, V.M. Plasma Electron Temperatures and Electron Energy Distributions Measured by Trace Rare Gases Optical Emission Spectroscopy. *J. Phys. D. Appl. Phys.* **2004**, *37*, R217. [CrossRef]
3. Boffard, J.B.; Lin, C.C.; DeJoseph, C.A. Application of Excitation Cross Sections to Optical Plasma Diagnostics. *J. Phys. D. Appl. Phys.* **2004**, *37*, R143. [CrossRef]
4. Soltani, B.; Habibi, M. Development of a Helicon Plasma Source for Neutral Beam Injection System of the Alborz Tokamak. *J. Fusion Energy* **2017**, *36*, 152–160. [CrossRef]
5. Afsharmanesh, M.; Habibi, M. Development of a Helicon Ion Source: Simulations and Preliminary Experiments. *Rev. Sci. Instrum.* **2018**, *89*, 033301. [CrossRef]
6. Crolly, G.; Oechsner, H. Comparative Determination of the Electron Temperature in Ar- and N2-Plasmas with Electrostatic Probes, Optical Emission Spectroscopy OES and Energy Dispersive Mass Spectrometry EDMS. *EPJ Appl. Phys.* **2001**, *15*, 49–56. [CrossRef]
7. Evdokimov, K.E.; Konischev, M.E.; Pichugin, V.F.; Sun, Z. Study of Argon Ions Density and Electron Temperature and Density in Magnetron Plasma by Optical Emission Spectroscopy and Collisional-Radiative Model. *Resour. Technol.* **2017**, *3*, 187–193. [CrossRef]
8. Shinohara, S. Helicon High-Density Plasma Sources: Physics and Applications. *Adv. Phys. X* **2018**, *3*, 1420424. [CrossRef]
9. Chen, F.F. Helicon Discharges and Sources: A Review. *Plasma Sources Sci. Technol.* **2015**, *24*, 014001. [CrossRef]
10. Degeling, A.W.; Boswell, R.W. Modeling Ionization by Helicon Waves. *Phys. Plasmas* **1997**, *4*, 2748–2755. [CrossRef]
11. Simonin, A.; Achard, J.; Achkasov, K.; Bechu, S.; Baudouin, C.; Baulaigue, O.; Blondel, C.; Boeuf, J.P.; Bresteau, D.; Cartry, G.; et al. R&D around a Photoneutralizer-Based NBI System (Siphore) in View of a DEMO Tokamak Steady State Fusion Reactor. *Nucl. Fusion* **2015**, *55*, 123020. [CrossRef]
12. Sharma, N.; Chakraborty, M.; Neog, N.K.; Bandyopadhyay, M. Design of a Helicon Plasma Source for Ion–Ion Plasma Production. *Fusion Eng. Des.* **2017**, *117*, 30–38. [CrossRef]
13. Hellsten, T.; Johnson, T.; Sharapov, S.E.; Kiptily, V.; Eriksson, J.; Mantsinen, M.; Schneider, M.; Rimini, F.; Tsalas, M. RF Heating for Fusion Product Studies. *AIP Conf. Proc.* **2015**, *1689*, 060007. [CrossRef]
14. Chen, F.F.; Torreblanca, H. Permanent-Magnet Helicon Discharge Array. *IEEE Trans. Plasma Sci.* **2011**, *39*, 2452–2453. [CrossRef]
15. Fang, T.Z.; Wang, L.; Jiang, D.M.; Zhang, H.X. Helicon Discharge Using a Nagoya Type III Antenna. *Chin. Phys. Lett.* **2001**, *18*, 1098–1100. [CrossRef]
16. Yoon, O.J.; Kang, S.M.; Moon, S.M.; Jung, J.K.; Choi, I.S.; Yoon, K.R. Deposition of Iron Nanoparticles onto Multiwalled Carbon Nanotubes by Helicon Plasma-Enhanced, Chemical Vapor Deposition. *J. Non Cryst. Solids* **2007**, *353*, 1208–1211. [CrossRef]
17. Krishnamurthy, A.; Chen, G.; Ulmen, B.A.; Keutelian, P.; Orcutt, J.; Miley, G.H. Helicon Injected Inertial Plasma Electrostatic Rocket, HIIPER. *Plasma Phys. Control. Fusion* **2012**, *33*, 1–2.
18. Batishchev, O.V. Minihelicon Plasma Thruster. *IEEE Trans. Plasma Sci.* **2009**, *37*, 1563–1571. [CrossRef]
19. Charles, C. Plasmas for Spacecraft Propulsion. *J. Phys. D. Appl. Phys.* **2009**, *42*, 163001. [CrossRef]
20. Buttenschoen, B.; Fahrenkamp, N.; Grulke, O. A High Power, High Density Helicon Discharge for the Plasma Wakefield Accelerator Experiment AWAKE. *Plasma Phys. Control. Fusion* **2018**, *60*, 075005. [CrossRef]
21. Pandey, A.; Sudhir, D.; Bandyopadhyay, M.; Chakraborty, A. Conceptual Design of a Permanent Ring Magnet Based Helicon Plasma Source Module Intended to Be Used in a Large Size Fusion Grade Ion Source. *Fusion Eng. Des.* **2016**, *103*, 1–7. [CrossRef]

22. Briefi, S.; Fantz, U. Investigation of Helicon Discharges as RF Coupling Concept of Negative Hydrogen Ion Sources. *AIP Conf. Proc.* **2013**, *1515*, 278–283. [CrossRef]
23. Simonin, A.; Agnello, R.; Bechu, S.; Bernard, J.M.; Blondel, C.; Boeuf, J.P.; Bresteau, D.; Cartry, G.; Chaibi, W.; Drag, C.; et al. Negative Ion Source Development for a Photoneutralization Based Neutral Beam System for Future Fusion Reactors. *New J. Phys.* **2016**, *18*, 125005. [CrossRef]
24. Agostinetti, P.; Franke, T.; Fantz, U.; Hopf, C.; Mantel, N.; Tran, M.Q. RAMI Evaluation of the Beam Source for the DEMO Neutral Beam Injectors. *Fusion Eng. Des.* **2020**, *159*, 111628. [CrossRef]
25. Choe, K.; Jung, B.; Chung, K.J.; Hwang, Y.S. Development of a Radio Frequency Ion Source with Multi-Helicon Plasma Injectors for Neutral Beam Injection System of Versatile Experiment Spherical Torus. *Rev. Sci. Instrum.* **2014**, *85*, 11–15. [CrossRef]
26. Chen, F.F.; Evans, J.D.; Tynan, G.R. Design and Performance of Distributed Helicon Sources. *Plasma Sources Sci. Technol.* **2001**, *10*, 236–249. [CrossRef]
27. Bartschat, K.; Kushner, M.J. Electron Collisions with Atoms, Ions, Molecules, and Surfaces: Fundamental Science Empowering Advances in Technology. *Proc. Natl. Acad. Sci. USA* **2016**, *113*, 7026–7034. [CrossRef]
28. Pitchford, L.C.; Alves, L.L.; Bartschat, K.; Biagi, S.F.; Bordage, M.C.; Phelps, A.V.; Ferreira, C.M.; Hagelaar, G.J.M.; Morgan, W.L.; Pancheshnyi, S.; et al. Comparisons of Sets of Electron-Neutral Scattering Cross Sections and Swarm Parameters in Noble Gases: I. Argon. *J. Phys. D. Appl. Phys.* **2013**, *46*, 334001. [CrossRef]
29. Chutjian, A.; Cartwright, D.C. Electron-Impact Excitation of Electronic States in Argon at Incident Energies between 16 and 100 EV. *Phys. Rev. A* **1981**, *23*, 2178–2193. [CrossRef]
30. Chilton, J.E.; Lin, C.C. Measurement of Electron-Impact Excitation into the 3p53d and 3p55s Levels of Argon Using Fourier-Transform Spectroscopy. *Phys. Rev. A* **1999**, *60*, 3712–3721. [CrossRef]
31. Bartschat, K.; Zeman, V. Electron-Impact Excitation from the (3p54s) Metastable States of Argon. *Phys. Rev. A* **1999**, *59*, R2552–R2554. [CrossRef]
32. Maloney, C.M.; Peacher, J.L.; Bartschat, K.; Madison, D.H. Excitation of Ar 3p54s-3p54p Transitions by Electron Impact. *Phys. Rev. A* **2000**, *61*, 022701. [CrossRef]
33. Zatsarinny, O.; Wang, Y.; Bartschat, K. Electron-Impact Excitation of Argon at Intermediate Energies. *Phys. Rev. A* **2014**, *89*, 022706. [CrossRef]
34. Jönsson, P.; Gaigalas, G.; Bieroń, J.; Fischer, C.F.; Grant, I.P. New Version: Grasp2K Relativistic Atomic Structure Package. *Comput. Phys. Commun.* **2013**, *184*, 2197–2203. [CrossRef]
35. Srivastava, R.; Stauffer, A.D.; Sharma, L. Excitation of the Metastable States of the Noble Gases. *Phys. Rev. A* **2006**, *74*, 012715. [CrossRef]
36. Gangwar, R.K.; Sharma, L.; Srivastava, R.; Stauffer, A.D. Argon Plasma Modeling with Detailed Fine-Structure Cross Sections. *J. Appl. Phys.* **2012**, *111*, 053307. [CrossRef]
37. Gangwar, R.K.; Sharma, L.; Srivastava, R.; Stauffer, A.D. Electron-Impact Excitation of Argon: Cross Sections of Interest in Plasma Modeling. *Phys. Rev. A* **2010**, *81*, 052707. [CrossRef]
38. Dipti; Srivastava, R. Electron-Impact Excitation Rate-Coefficients and Polarization of Subsequent Emission for Ar+ Ion. *J. Quant. Spectrosc. Radiat. Transf.* **2016**, *176*, 12–23. [CrossRef]
39. Boffard, J.B.; Piech, G.A.; Gehrke, M.F.; Anderson, L.W.; Lin, C.C. Measurement of Electron-Impact Excitation Cross Sections out of Metastable Levels of Argon and Comparison with Ground-State Excitation. *Phys. Rev. A* **1999**, *59*, 2749–2763. [CrossRef]
40. Priti; Gangwar, R.K.; Srivastava, R. Collisional Radiative Model for Ar-O2 Mixture Plasma with Fully Relativistic Fine Structure Cross Sections. *Phys. Plasmas* **2018**, *25*, 043517. [CrossRef]
41. Dipti; Gangwar, R.K.; Srivastava, R.; Stauffer, A.D. Collisional-Radiative Model for Non-Maxwellian Inductively Coupled Argon Plasmas Using Detailed Fine-Structure Relativistic Distorted-Wave Cross Sections. *Eur. Phys. J. D* **2013**, *67*, 203. [CrossRef]
42. Shukla, N.; Gangwar, R.K.; Srivastava, R. Diagnostic of Ar-CO2 Mixture Plasma Using a Fine-Structure Resolved Collisional Radiative Model. *Spectrochim. Acta Part B At. Spectrosc.* **2021**, *175*, 106019. [CrossRef]
43. Gangwar, R.K.; Dipti; Srivastava, R.; Stafford, L. Spectroscopic Diagnostics of Low-Pressure Inductively Coupled Kr Plasma Using a Collisional–Radiative Model with Fully Relativistic Cross Sections. *Plasma Sources Sci. Technol.* **2016**, *25*, 035025. [CrossRef]
44. Hartgers, A.; van Dijk, J.; Jonkers, J.; van der Mullen, J.A.M. CRModel: A General Collisional Radiative Modeling Code. *Comput. Phys. Commun.* **2001**, *135*, 199. [CrossRef]
45. Wubetu, G.A.; Kelly, T.J.; Hayden, P.; Fiedorowicz, H.; Skrzeczanowski, W.; Costello, J.T. Recombination Contributions to the Anisotropic Emission from a Laser Produced Copper Plasma. *J. Phys. B At. Mol. Opt. Phys.* **2020**, *53*, 065701. [CrossRef]
46. Colombant, D.; Tonon, G.F. X-Ray Emission in Laser-Produced Plasmas. *J. Appl. Phys.* **2019**, *44*, 3524. [CrossRef]
47. Lu, H.; Su, M.; Min, Q.; Cao, S.; He, S.; Dong, C.; Fu, Y. Effect of Dielectronic Recombination on Charge-State Distribution in Laser-Produced Plasma Based on Steady-State Collisional-Radiative Models. *Plasma Sci. Technol.* **2020**, *22*, 105001. [CrossRef]
48. Belmonte, T.; Noël, C.; Gries, T.; Martin, J.; Henrion, G. Theoretical Background of Optical Emission Spectroscopy for Analysis of Atmospheric Pressure Plasmas. *Plasma Sources Sci. Technol.* **2015**, *24*, 064003. [CrossRef]
49. Mewe, R. On the Positive Column of Discharges in Helium at Intermediate Pressures. *Physica* **1970**, *47*, 373–397. [CrossRef]
50. Kramida, A.; Ralchenko, Y.; Reader, J.; Team, N.A. NIST Atomic Spectra Database (Version 5.8). Available online: Https://Physics.Nist.Gov/Asd (accessed on 25 August 2022).

51. Ali, M.A.; Stone, P.M. Electron Impact Ionization of Metastable Rare Gases: He, Ne and Ar. *Int. J. Mass Spectrom.* **2008**, *271*, 51–57. [CrossRef]
52. Straub, H.C.; Renault, P.; Lindsay, B.G.; Smith, K.A.; Stebbings, R.F. Absolute Partial and Total Cross Sections for Electron-Impact Ionization of Argon from Threshold to 1000 EV. *Phys. Rev. A* **1995**, *52*, 1115–1124. [CrossRef]
53. Kolts, J.H.; Setster, D.W. Decay Rates of Ar(4s,3 P 2), Ar(4s,3 P 0), Kr(5s,3 P 2), and Xe(6s,3 P 2) Atoms in Argon. *J. Chem. Phys.* **1978**, *68*, 4848. [CrossRef]
54. Yamada, I.; Ohtani, S.; Tawara, H.; Danjo, A.; Hirayama, T.; Matsumoto, A.; Suzuki, H.; Takayanagi, T.; Wakiya, K.; Yoshino, M. Electron Impact Ionization of F^+ Ne^+ Cl^+ and Ar^+ Ions. *J. Phys. Soc. Japan* **1989**, *58*, 3151–3156. [CrossRef]
55. Cui, R.; Han, R.; Yang, K.; Zhu, W.; Wang, Y.; Zhang, Z.; Ouyang, J. Diagnosis of Helicon Plasma by Local OES. *Plasma Sources Sci. Technol.* **2020**, *29*, 15018. [CrossRef]
56. Czerwiec, T.; Graves, D.B. Mode Transitions in Low Pressure Rare Gas Cylindrical ICP Discharge Studied by Optical Emission Spectroscopy. *J. Phys. D. Appl. Phys.* **2004**, *37*, 2827–2840. [CrossRef]

atoms

MDPI

Article

Electron Impact Ionization of Adenine: Partial Cross Sections

Mohammad Atiqur Rehman [1] and E. Krishnakumar [2,*]

[1] Tata Institute of Fundamental Research, Homi Bhabha Road, Colaba, Mumbai 400005, India
[2] Raman Research Institute, C V Raman Avenue, Sadashiva Nagar, Bangalore 560080, India
* Correspondence: ekkumar@tifr.res.in

Abstract: Electron ionization of a genetically important nucleobase, adenine, was investigated from threshold to 500 eV using crossed electron beam–effusive molecular beam geometry and time-of-flight mass spectrometry. We measured the complete set of absolute partial cross sections for adenine using the relative flow technique (RFT) up to an electron energy of 500 eV. Normalization to absolute values was performed using electron ionization cross sections for argon and the vapor pressure data of adenine. The total cross sections obtained by summing the partial cross sections were compared with the existing theoretical and experimental data. The appearance energies of various fragment ions were also measured and compared with the reported data. The prominence of ions with mass $(HCN)_n{}^+$ (n = 1 to 5) indicated a possible pathway to form adenine in the interstellar medium through aggregation of HCN units. Analysis of the partial cross sections for various groups of fragment ions as a function of electron energy was found to give insights into their composition.

Keywords: electron ionization; absolute cross sections; molecular processes—ISM: molecules

Citation: Rehman, M.A.; Krishnakumar, E. Electron Impact Ionization of Adenine: Partial Cross Sections. *Atoms* **2022**, *10*, 100. https://doi.org/10.3390/atoms10040100

Academic Editor: David D. Reid

Received: 30 August 2022
Accepted: 18 September 2022
Published: 23 September 2022

1. Introduction

Measurement of absolute total and partial cross sections for electron-impact ionization of biological molecules is important in radiation biology. Any high energy radiation on interaction with matter in a condensed form leaves a final trail of low energy electrons [1,2]. These electrons are produced through a cascade of ionization processes. Thus, the large number of secondary electrons and ions produced in the process carry a dominant fraction of the energy of the incident radiation [2,3]. In biological matter, these charge particles can interact resonantly or directly with the biomolecules through a series of reactions, causing damage to the DNA and the RNA in terms of either single or double strand breaks [3–5]. The direct interaction can break the backbone of the DNA, while the resonances or transient anion formation will create neutral radicals and anionic fragments [1–5]. Thus, to understand the radiation damage and its complete description, the entire sequence of events leading to the final chemical state of the molecules must be known, and the mechanisms involved must be understood. The complete set of absolute cross sections resulting from low to intermediate energy electron collisions with DNA molecules and its building blocks are needed as input in bio-chemical models as well as in Monte Carlo particle track simulations used to study damage in living cells induced by ionizing radiations, nano- and micro-dosimetry, and cancer therapy [6–9]. Monte Carlo track simulations [6] of radiation damage accounts for ionization, but the probability of simultaneous ionization and dissociation, known as dissociative ionization, has not been incorporated in these simulations or in any other model due to the unavailability of either theoretical or measured data on such processes for DNA bases. These data are in the form of partial cross sections, which are the cross sections for the formation of an ion of specific mass-to-charge ratio. Despite its importance, very limited data on the absolute partial ionization cross sections of the DNA bases exist though absolute total ion cross sections for the DNA and RNA bases up to 200 eV have been reported [10–13]. In recent works on adenine ionization, Minaev et al. [11] studied the formation of positive and negative ions of adenine under the impact of electrons (from 0.1 to 200 eV) using the crossed electron and molecular beam technique. The method measures the molecular beam intensity and determines the total cross sections for the formation of positive and negative ions of the studied molecules, their energy dependences, and absolute values. Quantum chemical calculations are performed for the studied molecules, ions, and fragments for interpretation of the crossed beams experiments. Jochims et al. [14] used photoionization mass spectrometry with synchrotron radiation in the 6–22 eV photon energy range to investigate fragmentation pathways, ionization energies, and ion appearance energies (AEs) and compared them with the results of electron impact. Dawley et al. [15] investigated the electron ionization of adenine near the threshold region using a high-resolution hemispherical electron monochromator and a quadrupole mass spectrometer. Ion efficiency curves of the threshold regions and the corresponding appearance energies (AEs) are presented for the parent cations and the five most abundant fragment cations of each molecule. The enthalpies of the possible reactions from the adenine were also obtained computationally, and ionization energies were calculated using quantum chemical calculations. van der Burgt et al. [16] measured the mass spectra, the relative ion yield, and appearance energies for various fragment ions up to 100 eV. They obtained the partial cross sections after normalizing their data to the theoretical values of total ion cross sections. Very recently, Ostroverkh et al. [17] measured the mass spectrum at 70 eV in a crossed beam experiment and the ion yield curves near the threshold.

On the theoretical side, there have been several reports on the total ionization cross sections. These include those using semi-classical Deutsch-Märk formalism and Binary-Encounter Bethe (BEB) formalism by Bernhardt and Paretzke [18] and Możejko et al. [19], Peudon et al. [20], and Bull et al. [21] using BEB formalism. Huo et al. provided data using an improved binary-encounter dipole (iBED) model [22], and Vinodkumar et al. [23] used a Spherical Complex Optical Potential (SCOP) model and scattering theory. Champion [24] used the Born approximation to calculate the cross sections. The most recent study was by Tan et al. [25], using a semi-empirical approach. The electron ionization induced fragmentation of adenine was studied by Bauer and Grimm [26] using semi-empirical and density functional theory.

Adenine, a purine nucleobase, has a significant role in both protein synthesis and cellular respiration because it is a main component of DNA and ATP. Additionally, adenine is interesting to the astrobiology and astrochemistry community because of the possibility of its formation in space [27–29] and its potential role in the synthesis of larger bio-complexes [30]. Adenine has been found in meteoritic materials [31,32], and a possible precursor of adenine, cyanomethanimine, has been detected in the interstellar medium [33]. Meteorites provide a record of the chemical processes that occurred in the solar system before life began on Earth. Several organic molecular species have been identified so far in astronomical environments, containing the main functional groups necessary to initiate a complex organic chemistry and indicating that many more complex molecules are synthesized in space [34–36]. Until recently, the role of HCN in forming prebiotic molecules in solutions has been speculated [37,38]. However, Chakraborti et al. [27] proposed, on the basis of model calculations, that adenine can be produced in space by HCN oligomerization in the gas phase. Therefore adenine can be viewed as a pentamer of HCN, formed by successive addition of HCN molecules in four steps [39] in gas phase reactions in the dense interstellar clouds in star-forming regions.

Considering the paucity of accurate data, we made extensive measurements on the electron ionization cross sections for DNA bases; total ion cross sections have already been communicated [40]. The only reports on partial cross sections have been by Minaev et al. at 95 eV [11] and by van der Burgt et al. up to 100 eV [16]. Minaev et al. derived the partial cross sections from the total ionization cross sections measured by them, while van der Burgt et al. derived them from theoretical total ionization cross sections. Here, we present the partial ion cross sections by a direct measurement using the relative flow technique up to the electron energy of 500 eV. We also report ionization energy (IE) for the parent ions and appearance energies (AE) for most fragment ions.

2. Experiment

2.1. Measurement

Accurate determination of target density and the electron current in the interaction volume is needed in a crossed beam experiment to measure the absolute cross sections. It is difficult to determine the target density profile in a beam and its exact volume overlap with the electron beam in the cross-beam experimental set up. This is overcome by a normalization technique called the Relative Flow Technique (RFT) [41], which compares the intensity of the sample species with that of a standard species of known cross section, provided that the measurements for both the gases are carried out under identical experimental conditions. This is achieved by the gas flowing through a capillary under a molecular flow regime so that the molecules effusing out of it will have a specific beam profile independent of the nature of the molecules, and hence the geometry of the interaction volume becomes independent of the nature of the gas. The only change will be a constant multiplier, which depends on the pressure behind the capillary and can be measured accurately. While this technique is rather straightforward for molecules that have enough vapor pressure at room temperature, those with low vapor pressure, especially those that are solids at room temperature, need to be heated to elevated temperatures to increase their vapor pressure. Pressure measurements at such elevated temperatures are technically a very difficult task in cross-beam experimental setups due to the absence of appropriate manometers. The measurements of adenine is one such case. We have overcome this problem by measuring the temperature accurately and using the temperature versus vapor pressure data.

Though the experiment has been described elsewhere [42,43], for completeness we describe the essential features here. It uses an effusive molecular beam formed by a capillary, a magnetically collimated and pulsed electron gun, a Faraday cup to measure the incident electron current, a Time of Flight Mass Spectrometer (ToFMS) to mass select the ions, a pair of micro-channel plates (MCP) in chevron configuration as a detector, and the associated electronics and computer program to record the ion signal as a function of the mass and electron energy.

The effusive beam of adenine molecules is prepared by heating the commercially available sample (Sigma-Aldrich, St. Louis, MO, USA) in the oven to effuse through the capillary directly into the interaction region, as described earlier [43]. The temperature of the oven was gradually increased over a period of a few days to the required value of 474 K, while monitoring the mass spectrum of emanated water vapor. This was to ensure the uniform heating of the sample to thermal equilibrium as well as eliminating the water vapor from the sample. The uniformity of the temperature along the sample was ensured by monitoring it close to both the ends by different well-calibrated thermocouples. The temperature of the sample was maintained at 474 K to obtain reasonable target densities but at a low enough pressure to ensure molecular flow regime. At this temperature, the calculated pressure was 45.82 mTorr, low enough to ensure molecular flow through the capillary (0.2 mm diameter) so that the ratio of mean free path to the capillary diameter was much higher than unity.

The pulsed electron beam was operated at a repetition rate of 5 kHz, with the pulse duration being 300 ns. A pulsed extraction field was applied immediately after the electron pulse to extract the cations resulting from the electron–molecule collisions. The ions were detected by the MCP detector mounted at the end of the flight tube and operated in the pulse counting mode. The ion extraction field and the ion optics, including the flight tube, and the detector biases were optimized to ensure no discrimination in the collection, transmission, and detection of the ions due to the initial velocity distribution of the ions and their mass-to-charge ratio. Uniform detection efficiency was ensured by changing the bias voltage combination on the front and back of the MCP detector plates and looking for relative variation of the intensity of the highest mass peak to that one below 40 amu/e, and the operating voltages were fixed in the range where this ratio showed saturation. The uniformity of the detector efficiency up to mass-to-charge ratio of 130 amu/e was also confirmed by measuring the positive ion counts on electron impact from Ar and Xe and comparing the ratios of Ar^{2+} to Ar^{+}, Xe^{2+} to Xe^{+}, and Xe^{3+} to Xe^{+} with those reported in the literature. These were found to be in agreement within the experimental uncertainties.

A computer-controlled programming system allowed the storage of the mass spectra, the electron current as a function of the electron energy, and the ion yield curves. The ion yields curves were converted to absolute cross sections by using the Ar^{+} cross section from Argon at 100 eV as the standard. The measurements for Ar were carried out by flowing the gas through the same oven and the capillary tube used for making measurements on the sample molecules. Argon measurements were done immediately before and after under similar experimental conditions, except for the oven temperature. During Ar measurements, the oven was kept at room temperature in order to prevent interference from the adenine sample. An independent set of measurements on argon were carried out at different temperatures to determine any systematic error arising due to higher temperatures. These measurements showed no effect on the argon data with temperature. The pressure of Ar behind the capillary was measured using a capacitance manometer, while the pressure for the sample molecules was determined by using the vapor pressure data of adenine reported by W. Zielenkiewicz [44]. The vapor pressure for adenine at 474 K was determined from the relation [44] given as follows:

$$\log\left(\frac{p}{p0}\right) = 38.4 \pm 0.6 - \frac{17350 \pm 252}{T}$$

In RFT, under identical conditions, we measured the intensities, N_u, of an ion u of the sample gas under study and N_s of an ion of known cross sections, which was used as a standard (s). The partial ionization cross section σ_u can be related with known cross sections (σ_s), as

$$\sigma_u = \sigma_s \times \frac{N_u}{N_s} \times \frac{I_s}{I_u} \times \frac{F_s}{F_u} \times \sqrt{\frac{M_s}{M_u}} \times \frac{K_s}{K_u}$$

where N represents the intensity of each ion, F is the flow rate of individual gases, M is the molecular weight of each gas, and I is the time-averaged electron beam current. K_s and K_u are the efficiency of collection, transmission, and detection of the ion used as the standard and the one under measurement, respectively. This equation can be further simplified, since $F.M^{1/2}$ is proportional to pressure P behind the capillary under molecular flow conditions, as

$$\sigma_u = \sigma_s \times \frac{N_u}{N_s} \times \frac{I_s}{I_u} \times \frac{P_s}{P_u} \times \frac{K_s}{K_u} \quad (1)$$

One of the crucial aspects of the experiment, which was described above, is to ensure that K is the same for all ions so that the ratio K_s/K_u is unity. The possibility of thermal decomposition of adenine was ruled out by measuring the mass spectra at fixed electron energy as a function of temperature over the range of temperature. No change in the relative intensity of the mass spectra or no new fragments were observed with a change in temperature. The possible thermal decomposition of the sample could also be identified by the change of color of the powder on visual inspection of the remaining sample after the experiment. By taking into account all these, we ensured that thermal decomposition did not contribute any erroneous signal in our measurements.

In the measurement procedure, to begin with, the mass spectrum was measured in crossed beam mode at an electron energy of 100 eV and a temperature of 474 K. Next, the ion yield curves for all the fragments were recorded from 0–500 eV. Then, argon was introduced, and Ar^{+} counts were recorded at 100 eV. During the crossed beam measurements the contribution to ion counts coming from the uniformly filled background molecules due to scattering at the surfaces was subtracted from the measured crossed beam data in order to have the contribution from the beam alone. This was done by placing the oven in such a way that no molecular beam was available in the interaction region and the chamber was filled uniformly by the sample gas. The background mass spectrum was subtracted, and the resultant ion counts were normalized to the electron current and pressure to obtain the normalized counts for each mass fragment. The mass scale was calibrated using Ar^+ and Ar^{++}. Partial cross sections were calculated using Equation (1) given above.

In order to ensure complete collection and detection of the ions, we used voltage biases on the ToF spectrometer, which did not provide the best mass resolution. This resulted in mass peaks overlapping with each other in certain mass ranges, forming a few envelopes. Individual contributions for ions of a given mass-to-charge ratio (m/z) were obtained by unfolding each envelope to individual peaks, assuming Gaussian shapes. The area under each Gaussian was taken as the contribution due to that particular ion.

2.2. Uncertainty Estimation

The uncertainties in the present measurements are estimated as follows. The uncertainties arise from the relative cross section (ion yield curve) measurement as well as from the normalization to the absolute numbers. The uncertainty in the ion yield curves arise mainly from possible overlap volume change (in the electron beam and molecular beam) with respect to the change in electron energy apart from possible variations in the electron current source and target pressure. While strict control could be made in the current source variation and target pressure variation, the systematic error from the volume overlap change is difficult to control. The uncertainty due to temporal variation in the experimental parameters is reduced by completing one scan of the entire electron energy up to 500 eV within 300 s. The reproducibility of the ion yield curves is within 1% over the entire energy range. The volume overlap changes with the electron beam is checked by making measurements on Ar+ from Ar in the electron energy range used here and was consistent with existing data [45] within 5%. Wherever the mass peaks were not clearly resolved, we employ deconvolution, leading to uncertainty in the relative intensities of the peaks thus obtained. The contribution due to this varies depending on the overall shape of the envelopes of peaks and the statistics. In addition to this, uncertainties also arise from measurement of electron current, pressure and statistical errors from ion counting. Non-linear effects due to pulse pile up are addressed by keeping the maximum total count rate (inclusive of all the masses) at the most at one-tenth of the electron beam pulse rate, even though we use a multi-hit card for data acquisition. It is assured that the temperature remains the same at the oven and other end of capillary within 0.1 °C, which causes an uncertainty of less than 0.1% in the vapor pressure. The uncertainty quoted in [44] in the temperature dependence of the vapor pressure data is about 2%, which we assume as the uncertainty in the pressure of adenine. The uncertainty in the Ar pressure measurement is about 0.1%. One important source of uncertainty in the absolute cross sections arises from that of the Ar ionization cross sections we use for normalization. We use the most recent data [45], which quotes an uncertainty of 5%. These data are consistent with other measurements [46–48]. An analysis of the data from all the four sets [45–48] gives a standard deviation of 2.9%. Assuming 5% uncertainty in the Ar data and combining it with the uncertainty in the counting statistics, including that arising from the Gaussian deconvolution of the peaks, pressure measurements and the uncertainty in the ion yield curve provide a total uncertainty of 6% in the present measurements in absolute cross sections for all the ions, which are at least 10% of the intensity of the parent ion. The maximum uncertainty in the detection efficiency for ions over the range of interest is about 10%. This gives a total uncertainty in the cross section measurements of 12%.

3. Results and Discussion

3.1. Mass Spectrum

The relative intensities of the peaks measured at 70 eV and 100 eV electron impact is given in Table 1 along with other reported data. The dominant peaks in the spectrum are $(CNH)_n{}^+$ with n = 1 to 5 and $HCNH^+$. We could not collect H^+ and measure its intensity reliably due to the relatively large initial kinetic energy with which it is produced and also due to it getting deflected away during its flight to the detector by the magnetic field used to collimate the electron beam.

Table 1. Relative intensities of major fragment ions from adenine.

m/z	Electron Impact						Photon Impact
	Present Work 100 eV	Minaev et al. [11] 95 eV	Present Work 70 eV	NIST [50] 70 eV	Rice et al. [49] 70 eV	Ostroverkh et al. [17] 70 eV	Jochims et al. [14] 20 eV
136	8.0	4.29	8	7.6	-	7.86	-
135	100	100	100	100	100	100	100
134	6	2.86	6.4	3	3	3.32	10
120	0.5	-	0.44	-	3	1.55	1
119	1.3	-	1.3	1.2	-	1.44	3
118	0.8	-	0.81	0.9	-	1.11	-
108	25	35.7	23.7	27	34	30.3	57
107	1.9	5.0	1.8	1.9	3	2.44	10
92	2.4	-	2.4	1.5	-	1.55	9
83	1.3	-	1.2	0.8	-	0.441	-
82	2.0	-	1.9	0.6	-	0.332	-
81	10.2	22.9	9.9	9.7	19	10.9	50
80	5.3	10.0	5.1	3.3	7	4.43	10
70	5.0	5.0	5.0	3.6	5	4.42	17
66	11.8	20.0	11.8	6.3	15	7.59	41
65	3.9	7.14	3.6	1.8	6	2.58	0
56	2.3	-	2.3	1.5	-	1.37	-
55	6.9	11.4	6.6	2.4	-	2.33	-
54	25	32.9	23	11.2	31	15.3	55
53	24	25.7	21	7.6	24	8.45	28
43	7.6	10.0	7.3	4.3	12	6.542	34
42	2.6	5.0	2.3	0.5	3	0.443	16
41	2.7	2.85	2.6	1.1	2	1.03	7
40	6.7	10.0	5.8	1.6	5	1.89	1
39	10.4	10.0	8	2.2	8	2.58	1
38	9.9	11.4	7.2	2.4	10	2.44	0
29	16	8.57	15.5	3.6	12	7.58	60
28	82	21.4	74	18.5	78	67.1	110
27	12.8	10.0	7.7	1.9	12	0.551	10
18	-	-	-	-	-	11.7	-
17	-	-	-	-	-	4.20	-
15	-	-	-	-	-	4.43	-
14	-	-	-	-	-	3.43	-
13	3.9	-	1.5	-	-	-	-
12	2.8	-	0.85	-	-	-	-

The relative intensities measured by Rice and Dudek [49] matches well with our 100 eV electron impact mass spectra within experimental accuracies for most fragments, the notable exceptions being $m/z = 54$ and 27. The high prominence of the important ion fragment at $m/z = 28$ observed in the present measurement and in [14,49] is not observed in [11,50]. Recent measurements of van der Burgt [16] also observed relatively high intensity of $m/z = 28$. Dawley et al. [15] state that the relatively large intensity of $m/z = 28$ observed by them may due to contamination from N_2. They observed larger intensity for this ion than the parent ion. Considering this, we have tried to eliminate and evaluate the presence of N_2 in our experiment. We eliminate possible external leak from the atmosphere by having a background pressure of a few times 10^{-9} Torr in the experiment. Further, the analysis of the mass spectrum of adenine shows that if at all, there is very little presence of N_2 in our experiment. This is done by looking at the ratio of $m/z = 28$ to that $m/z = 14$ (N^+

and N_2^{2+}). The reported value of this ratio at 100 eV from N_2 is 3.2 [51], where as in the present measurements it is about 60. Even assuming that N^+ collection and detection after mass analysis may be skewed due to their relatively large kinetic energy, this ratio is too large. In fact, the measured ratio for N_2^+ to N_2^{2+} (both having thermal energies) from N_2 is about 53 [52]. Our experiment is designed to collect, mass analyze and detect all the ions independent of their initial velocity distribution. We also note the considerable intensity of $m/z = 29$ (about 20 % of that of $m/z = 28$), which can only be formed from adenine. Due to its composition, CH_3N^+, its formation needs rearrangement of at least one H atom from another site in the molecule. Even assuming a H atom scrambling may be common place in an excited adenine ion prior to its fragmentation, the probability for the formation of this ion is not too small. Hence the observed high cross section for $m/z = 28$, which arises from a direct fragmentation process, is not surprising. In view of these we believe that the contribution to $m/z = 28$ from N_2^+, if at all is very small.

We also observe the ion of $m/z = 67.5$ with reasonable intensity, which corresponds to doubly charged parent ion, $C_5H_5N_5^{2+}$. We could not obtain a reliable number for its intensity as it is mixed with $m/z = 67$ and 68 due to limited mass resolution. The ions of $m/z = 12$ and 13 are also seen in the present data. We note that the low mass ions have relatively higher ion yield in comparison to other experiments. The variation observed in different experiments in the relative intensity of the observed mass peaks may be due to systematic errors in the various experiments. There are two sources of systematic errors that may affect the observed relative ion intensity distribution in the mass spectra. The first one is the variation in efficiency of the detector as a function of the mass-to-charge ratio. The second one is the collection and transmission efficiency of the mass spectrometer. The collection and transmission depend on the initial kinetic energy and angular distribution of the ions. In addition, their mass-to-charge ratio may also come into effect when a quadrupole mass spectrometer is used for mass analysis. The lighter ions are likely to be affected more by the kinetic energy discrimination, while the heavier ions are likely to be affected by the detection efficiencies. In the present measurements, except for the case of H^+, we ensure that the uncertainties due to collection and detection efficiencies are minimized by using narrow electron gun pulses, a large pulsed field extraction and appropriate detector biases, and sensitive pulse counting electronics. Automation of the experiment allowed us to run the experiment without interruption for several days continuously in order to build up adequate statistics while using minimum electron beam current and target pressure. This eliminated various other systematic errors arising from deviation from single collision conditions and detector and counting electronics pile-up issues.

3.2. Appearance Energies and Fragmentation Channels

We measured ion yield curves for most fragment ions from adenine in the energy range 0–500 eV, and the curves near threshold for some prominent ions are shown in Figure 1. The changes in slope of the ion yield curves are indicated by solid lines superimposed on them in order to obtain the appearance energies (AEs). The solid lines were obtained using linear fit. We are unsure whether the exponential fit based on Wannier threshold law is applicable or is critical for determining the appearance energies in the case of ionization of poly-atomic molecules due to the absence of well-separated molecular ion states. This becomes particularly troublesome for fragment ion formation from them, as we lack information on the dissociating states and their dynamics. In view of this, we use the relatively simple approach of linear fit. The appearance potentials of the major m/z ions are listed in Table 2, as measured from the onsets in the ion yield curves. The present AE values in this table were derived by reading out the intersection of linear fit with the energy axis. Table 2 also includes the AE measured by Jochims et al. [14] using photon impact and Dawley et al. [15] using electron impact ionization. In most cases, we observed two AE values due to the clear change in the slope of the ion yield curve. Two slopes could be due to the particular ion being formed by two different pathways, with the second one contributing at higher energies. This kind of change of slope in ion yield curve is not apparent in the parent ion and $C_4H_4N_4^+$ fragment ion (Figure 1). In most ion fragments, our first threshold is in close agreement with the existing measurements using electron impact by Dawley et al. [15].

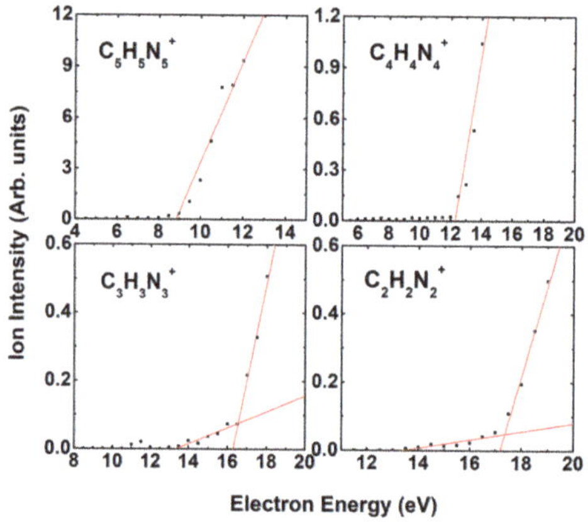

Figure 1. Appearance energies for some major fragment ions from adenine.

Table 2. Appearance energies of major ions from adenine in eV.

m/z	Photon Impact	Electron Impact				
	Jochims et al. [14]	Dawley et al. [15]	van der Burgt et al. [16]	Present		
				1st	2nd	Difference
135	8.2 ± 0.03	8.7 ± 0.3	8.0 ± 0.2	8.8 ± 0.3	-	-
108	11.56 ± 0.05	11.7 ± 0.2	11.3 ± 0.2	12.3 ± 0.5	-	-
94	-	-	-	15 ± 0.5	17 ± 0.5	2
93	-	-	-	15.2 ± 0.5	17.2 ± 0.5	2
92	-	-	-	15.6 ± 0.4	17.6 ± 0.4	2
91	-	-	-	16.5 ± 0.5	-	-
82	-	-	-	15.7 ± 0.4	18.2 ± 0.4	2.5
81	12.8 ± 0.1	14.14 ± 0.5	13.1 ± 0.2	13.5 ± 0.5	16.3 ± 0.5	2.8
80	-	15.1 ± 0.5	14.8 ± 0.7	17.5 ± 0.4	21.5 ± 0.4	4
70	13.1 ± 0.1	14.9 ± 0.2	12.6 ± 0.4	13.0 ± 0.4	16.5 ± 0.4	3.5
66	13.2 ± 0.1	14.2 ± 0.3	13.5 ± 0.2	16.5 ± 0.4	19.0 ± 0.4	2.5
65	-	17.9 ± 0.4	15.7 ± 1.3	17.0 ± 0.4	20.5 ± 0.4	3.5
64	-	-	-	18.5 ± 0.4	-	-
56	-	-	-	16.5 ± 0.4	18.6 ± 0.4	2.1
55	-	-	-	15.5 ± 0.5	18 ± 0.5	2.5
54	13.7 ± 0.1	14.6 ± 0.3	13.5 ± 0.3	13.5 ± 0.5	17.5 ± 0.4	4
53	-	16.7 ± 0.5	15.3 ± 0.5	16.0 ± 0.5	21 ± 0.4	5
51	-	-	-	14 ±0.5	21 ± 0.4	7
43	13.0 ± 0.1	14.0 ± 0.3	13.3 ± 0.6	13.0 ± 0.5	17.0 ± 0.4	4
42	-	-	-	14 ± 0.4	19 ± 0.4	5
41	-	-	-	16 ± 0.4	22 ± 0.4	6
40	-	15.7 ± 0.3	15.9 ± 0.2	16.0 ± 0.5	20.0 ± 0.5	4
39	-	18.1 ± 0.2	17.1 ± 0.5	14.5 ± 0.5	21.5 ± 0.4	7
38	-	-	-	15.0 ± 0.5	23 ± 0.5	8
29	14.0 ± 0.1	15.15 ± 0.15	13.7 ± 0.3	14.5 ± 0.5	-	-
28	13.1 ± 0.1	13.1 ± 0.5	12.9 ± 0.4	15 ± 0.4	17.0 ± 0.4	2

Table 2. *Cont.*

m/z	Photon Impact	Electron Impact				
	Jochims et al. [14]	Dawley et al. [15]	van der Burgt et al. [16]	Present		
				1st	2nd	Difference
27	-	13.5 ± 0.2	14.0 ± 0.6	14 ± 0.5	20 ± 0.5	6
25	-	-	-	14 ± 1	-	-
24	-	-	-	15 ± 1	-	-
15	-	-	-	12 ± 0.5	19 ± 0.5	7
14	-	-	-	19.5 ± 0.5	27 ± 0.5	7.5
13	-	-	-	26 ± 1	37 ± 1	11
12	-	-	-	27 ± 1	38 ± 1	11

For the parent ion, m/z = 135 amu, we measured an ionization energy of 8.8 ± 0.3 eV from the ion yield curve shown in Figure 1. This agrees with the 8.8 ± 0.2 eV of Minaev et al. [11] and 8.7 ± 0.3 eV of Dawley et al. [15] both using electron impact ionization and 8.6 ± 0.006 of Plutzer et al. [53] using resonance 2-photon ionization. Our measured value is higher than the 8.2 ± 0.03 of Jochims et al. [14] using photoionization and the vertical ionization energy of 8.08 eV and adiabatic ionization energy of 8.07 eV using quantum chemical calculations by Dawley et al. [15].

The experimental AE value for the formation of m/z = 108 ($C_4H_4N_4^+$) is 12.3 ± 0.5 eV in our work. This AE value matches with that of 12.3 ± 0.1 obtained by Pilling et al. [54] using the photoelectron-photoion coincidence technique (PEPICO) but higher than the 11.56 ± 0.05 eV determined in the photoionization study by Jochims et al. [14] and 11.7 ± 0.2 given by Dawley et al. [15]. Our measured AE value for the formation of m/z = 81 ($C_3H_3N_3^+$) is 13.50 ± 0.5 eV. This is in between the AE value of 14.14 ± 0.5 by Dawley et al. [15] using electron impact and 12.8 ± 0.1 by Jochims et al. using photon impact [14]. We observed an AE value of 13.5 ± 0.5 eV for the formation of m/z = 54 ($C_2H_2N_2^+$). This is close to the AE value of 13.7 + 0.1 by Jochims et al. [14] and lower than the 14.55 ± 0.3 by Dawley et al. [15]. For m/z = 28 ($HCNH^+$) we find the appearance energy to be 15 ± 0.4, which is slightly larger than the values obtained by Jochims [14] and Dawley et al. [15].

We find that a second threshold appears for the formation of fragment ions of m/z = 92 and below. These thresholds should be indicative of new production channels for the respective ions in which the corresponding neutral parts may be fragmenting further. It may also be due to electronic excitation in any of the fragments (ionic or neutral). Additional thresholds for several ions between m/z = 37 and 71 have been observed by van der Burgt et al. [16]. The differences in the first and second thresholds in the present measurements are shown in the last column in Table 2. We note that this difference is about 4 eV or less as we go down in mass, until the production of m/z = 53, which corresponds to $C_2HN_2^+$. The next lower mass we observe is 43, corresponding to $CH_3N_2^+$, which is structurally one less C atom and corresponds to further cleaving of the ring structure. From here on, the difference in the two thresholds increases and reaches as much as 11 eV for C^+ formation, with the notable exception of that of m/z = 28, which is CH_2N^+. This large difference for the lighter ions may also arise from the double ionization process. Since the neutral radicals also play a major role in the chemical reactions in the tracks of high energy radiation, the knowledge of the neutral radical formation plays a crucial role in radiation chemistry. As the detection of neutral fragments are experimentally very difficult, the only method is to model the dissociative ionization process through quantum chemical calculations in order to incorporate these additional channels. In this respect, identification of the higher thresholds may be useful.

The formation mechanism of various ions has been discussed by a number of authors [14–17,26,49,55]. The recent study by Dawley et al. [15] combined experimental results with quantum chemical calculations. The most important aspect of these studies is the identification of the formation channels for $(CHN)_n^+$ with n = 1 to 4 by elimination of 1 to 4 HCN radicals from the parent $C_5H_5N_5^+$ ion, as shown in Figure 2.

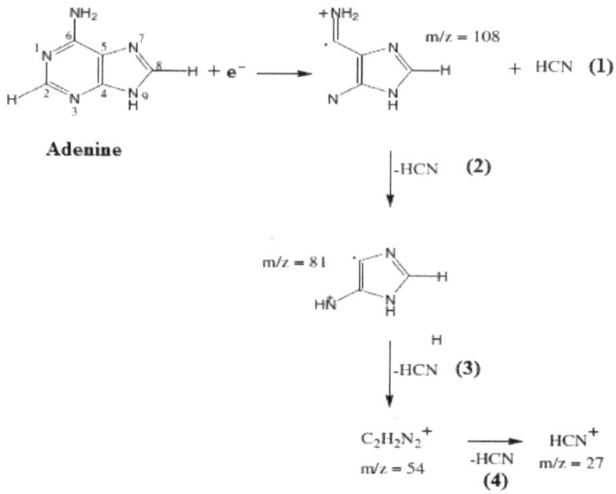

Figure 2. Principal pathway involved in fragmentation of adenine on ionization.

As shown in Table 1, we confirmed prominent fragment peaks corresponding to ions $C_nH_nN_n^+$ (n = 1 to 5) at m/z of 27 (CNH^+), 54 ($C_2N_2H_2^+$), 81 ($C_3N_3H_3^+$), 108 ($C_4N_4H_4^+$), and 135 ($C_5N_5H_5$+) in the mass spectrum of adenine. Each of these ions is formed by the loss of one HCN unit in succession from the adenine molecule. The ratios of partial ionization cross sections for these ions on elimination of HCN units relative to that of the parent ion at 31%, 11.2%, 23%, and 14%, respectively, at 100 eV are considerably high. This is consistent with photoionization [14]. The elimination of HCN is an important process in the electron ionization of polynitrogen heterocycles and constitutes the reaction sequence in many purine derivatives [56], including purine itself [57]. The decomposition of adenine following electron impact has been studied [49,55] on the basis of extensive isotopic substitution to establish the extent of site selectivity in fragment ion formation and by [14] using photon impact to show the successive expulsion of HCN from the adenine. Jochims et al. [14] also show that the corresponding ions $C_nH_nN_n^+$ (n = 1 to 5) have higher relative intensity both in electron and photon impact to adenine. It has been proposed that successive addition of HCN molecules in four steps [39] in gas phase reactions can produce adenine by oligomerization, where it can be seen as a pentamer of HCN [27]. In the fragmentation pattern of adenine, we see that the loss of each unit of HCN leads to a prominent ion with high partial cross sections. Successive loss of HCN is the most preferred pathway, and most ions are formed by losing HCN units in the dissociative ionization of adenine. This could be understood as a de-oligomeriztion of the adenine oligomer by electron impact.

3.3. Cross Sections

Partial cross sections for the formation of all the dominant ions are shown in Figures 3 and 4. In Figure 3, we give the cross sections for the most prominent ions, $(CHN)_n^+$ with n = 1 to 5 and CH_2N^+ (m/z = 28), along with the only other available data for comparison. The partial cross sections for a selected set of prominent ions along with the total cross sections are given in Table 3.

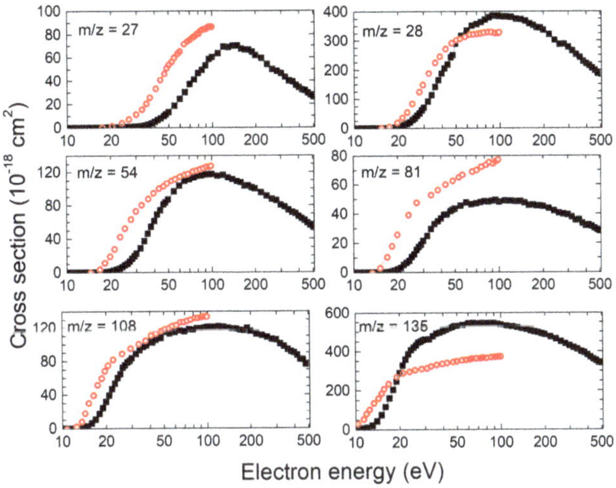

Figure 3. Partial cross sections for formation of $(CHN)_n^+$ (n = 1 to 5) and CH_2N^+. Black filled squares—present data and red open circles—from Ref. [16].

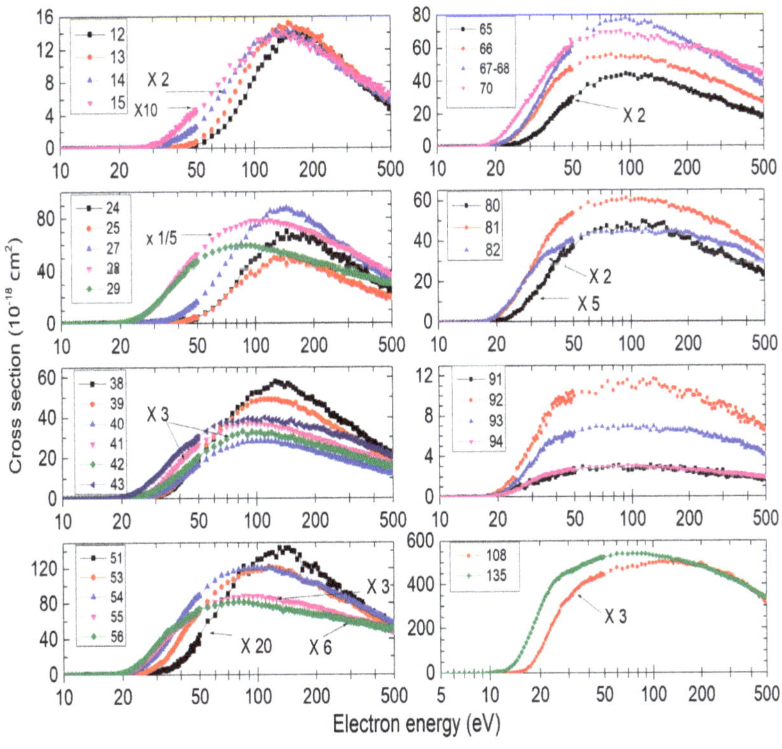

Figure 4. Partial ionization cross sections for all dominant ions from adenine. The numbers in the top left side of the panels represent the m/z values; 67–68 represent three ions of m/z = 67, 67.5, and 68, with 67.5 being the dominant one corresponding to doubly charged parent ion, $C_5H_5N_5^{2+}$.

Table 3. Partial ionization cross sections for prominent ions and the total ion cross section obtained by summing all the partial cross sections.

Electron Energy (eV)	Partial Ionization Cross Sections for Prominent Ions ($\times 10^{-18}$ cm^2)									Total Cross Section (10^{-16} cm^2)
	HCN^+	$HCNH^+$	C_2N^+	C_2NH^+	$C_2N_2H^+$	$C_2N_2H_2{}^+$	$C_3N_3H_3{}^+$	$C_4N_4H_4{}^+$	$C_5N_5H_5{}^+$	
8.5	0	0	0	0	0	0	0	0	0.20	0.002
9	0	0	0	0	0	0	0	0	0.34	0.003
10	0	0	0	0	0	0	0	0	2.3	0.03
11	0	0	0	0	0	0	0	0	7.8	0.105
12	0	0	0	0	0	0	0	0	9.3	0.138
13	0	0	0	0	0	0	0	0.026	21.7	0.138
14	0	0.025	0	0.015	0	0.01	0.03	0.22	42.9	0.221
15	0	0.05	0.039	0.018	0	0.013	0.04	1.0	74.6	0.407
16	0	0.10	0.070	0.022	0.03	0.02	0.07	2.4	116	0.694
17	0	0.20	0.070	0.032	0.045	0.054	0.22	5.9	159	1.09
18	0	0.73	0.093	0.044	0.12	0.20	0.51	10.4	203	1.53
19	0.079	1.5	0.11	0.059	0.11	0.50	1.2	17.3	246	2.05
20	0.13	3.1	0.23	0.071	0.15	0.97	2.1	24.0	288	2.60
25	0.37	22	0.57	0.25	2.0	8.3	10.4	59.7	418	3.16
30	0.92	65	1.1	1.3	9.1	22.3	21.4	80.3	459	5.93
35	1.8	122	2.9	4.1	22.9	41.4	30.3	93.0	483	8.21
40	4.8	178	7.6	9.2	39.4	61.0	36.7	100	501	10.8
45	8.4	225	12.6	15.7	56.3	77.3	41.1	105	520	12.5
50	13	260	16.9	20.9	68.1	86.8	43.2	107	529	14.4
55	19	289	22.5	26.7	78.7	96.4	45.3	111	535	15.5
60	24	316	26.1	30.5	88.5	103	47.5	115	544	16.9
65	30	334	31.1	35.4	94.7	109	46.3	113	544	18.0
70	37	351	34.2	37.9	99.6	111	46.9	114	542	18.6
75	41	356	37.4	40.2	103	113	47.7	116	545	19.2
80	47	364	40.0	43.1	106	115	48.3	115	543	19.6
85	49	371	41.1	45.3	110	116	48.2	118	544	19.9
90	53	379	44.0	46.7	111	117	48.6	118	544	20.1
95	59	384	46.2	48.5	112	117	49.2	119	540	20.6
100	60	385	46.6	48.8	113	117	48.0	119	536	21.0
125	68	380	51.5	49.4	116	114	47.8	120	527	20.6
150	70	369	49.9	47.8	112	109	47.4	120	512	20.6
175	68	353	45.7	44.8	100	102	45.2	117	495	19.4
200	62	341	42.2	42.4	98.4	99.6	44.8	120	485	19.1
225	58	321	41.0	38.2	90.4	94.1	42.8	113	469	17.0
250	53	305	37.1	35.6	85.5	90.0	41.4	111	455	16.6
275	48	289	33.3	33.5	80.3	84.1	40.1	108	443	16.2
300	46	279	32.6	30.9	79.8	80.1	39.2	105	428	15.3
325	41	264	30.6	28.8	72.4	76.8	37.8	101	418	14.7
350	39	249	28.3	27.4	68.2	73.0	35.2	97.5	405	13.9
375	36	239	25.0	25.4	63.8	69.7	33.6	93.5	391	13.4
400	34	224	23.6	24.3	60.0	66.3	31.8	89.7	381	12.9
425	32	215	22.6	22.6	56.4	64.0	30.8	86.9	368	12.5
450	29	197	20.8	20.2	52.8	58.6	29.2	81.9	346	11.8
475	28	193	20.3	19.8	52.3	57.8	29.0	80.8	347	11.1
500	26	178	18.3	18.9	47.7	54.2	28.1	77.2	340	11.5

From Figure 3, it is apparent that there is considerable difference between the existing data [16] and the present measurements. The difference exists both in magnitude and relative shape of the cross section curves (ion yield curves). The absolute magnitude given in [16] was obtained by normalization with theoretical total ion calculations and hence one may expect some difference. However, the difference in magnitude in the two data sets appears to be beyond this since the differences depend on m/z values. We are unable to attribute the reason for the observed differences except for a possible source of errors arising from ion transmission efficiencies of the mass spectrometer and detection efficiencies as a function of m/z. As explained in the experimental section, we have taken utmost care in eliminating these errors. The deviations in the shape of the cross section curves in the two measurements should not arise from these errors, at least in the zeroth order. A possible source of error in this respect seems to be overcorrection for the variation in the electron beam current as the current measured at the Faraday cup could be smaller than the actual current passing through the interaction volume. This effect is likely to be more pronounced as the electron energy decreases. As mentioned earlier, we have taken utmost care in eliminating this problem and characterized the experiment satisfactorily using known cross sections.

The complete set of partial ionization cross sections for all the ions of significant intensities are given in Figure 4. As expected, the cross sections for most of the ions rise relatively steeply toward the respective peaks and then decrease rather slowly towards higher energies. The energy corresponding to the cross section peak varies from ion to ion, but with a more or less systematic change as a function of m/z. For example, the peak for the parent ion ($m/z = 135$) occurs at an energy of about 75 eV, while that for C^+ occurs at about 160 eV. In the panels in Figure 4, we plot the cross sections for bunches of ions, irrespective of their absolute magnitudes, by using appropriate multipliers as given in the panels. This allows comparison of the relative shape of cross section curves within a given mass range, which may provide some insight into their formation process.

Among the ions with $m/z = 12$ to 15 (C^+, CH^+, N^+ and NH^+), NH^+ has the smallest appearance energy (AE) of 12 eV, followed by N^+ with an AE of 19.5 eV in comparison to CH^+ (AE = 26 eV) and C^+ (AE = 27 eV). That NH^+ has the least AE is not surprising, since it can be formed without breaking either of the two ring structures of adenine. However, it is interesting to note that in terms of absolute cross sections, the overall production of NH^+ is about a factor of 10 smaller than that of CH^+ and C^+ ions. Even the N^+ formation cross section is a factor of 5 higher than that of NH^+.

For the ions of m/z ranging from 24 to 29, we see three different groupings. $m/z = 28$ and 29 start off earlier than others, with AE of 15 and 14.5 eV respectively. Ion of $m/z = 27$ (HCN^+) has a yield curve different from others. The third group of $m/z = 24$ ($C_2{}^+$) and 25 (C_2H^+) seem to follow each other, but are different from the other two groups. C_2H^+ and $C_2{}^+$ have AEs of 14 eV and 15 eV, respectively. However, the ion yield curves are rather flat, with very small slope at low energies, and do not start rising till about 40 eV for C_2H^+ and 45 eV for $C_2{}^+$. HCN^+ has an AE of 18.5 eV and seems to have a second threshold for formation at 30 eV. Possible channels for formation of $m/z = 28$ ($HCNH^+$) have been discussed by Dawley et al. [15], and they do not favor direct formation through the fragmentation of the imidazole ring based on energetics. However, energy consideration may be valid at the threshold, and beyond the threshold there would be enough excess energy for the parent molecular ion to undergo any sort of structural change leading to fragmentation. The observed similarity in the yield curves of $HCNH^+$ to that of $HCNH_2{}^+$ (or H_2CNH^+) should help in identifying the fragmentation dynamics leading to these ions. That the $HCNH^+$ formation channel is the most dominant fragmentation channel of adenine ionization demands further study.

In the range of masses 38 to 43, we notice two groups with $m/z = 38$ and 39 having considerably different shapes for their ion yield curves in comparison to the other four. This difference could be attributed to their composition, which then would have a bearing on the fragmentation of the parent ion. The ions $m/z = 38$ and 39 would have compositions of C_2N^+ and C_2NH^+, respectively, while those of 40 to 43 would have a composition involving $CN_2H_n{}^+$ with n varying from 0 to 3. The difference in the number of C and N atoms in the two sets would have a dependence on how the ring structures in adenine break to form the respective fragments. All the ions in these groups of ions have two distinct AEs. As expected, in each of the groups, the AEs keep increasing with the reduction in the number of H atoms.

For the ions of m/z in the range of 51 to 56 also, we can clearly see the dependence of the ion yield curves on the composition of the ions in terms of the number of C, N, and H atoms. By comparing the shapes, one may even distinguish between two likely compositions of a given ion. In this set of ions, we notice three groupings: 51 (which could only be C_3NH^+) (first group), 53 ($C_3NH_3^+$ or $C_2N_2H^+$) and 54 ($C_2N_2H_2^+$) (second group) and 55 ($C_2N_2H_3^+$) and 56 ($C_2N_2H_4^+$) (third group). The ion of $m/z = 51$ is clearly demarcated from that of 54 with the distinction in the number of C and N atoms, and we see clear difference in the shape of the ion yield curves between the two. The ion of $m/z = 53$ could have in principle two compositions—$C_3NH_3^+$ or $C_2N_2H^+$. From a comparison of the curve of $m/z = 53$ with those of 51 and 54, one may conclude that $C_2N_2H^+$ is the most likely composition for 53. The difference between 53 and 54 at low energies and those of 55 and 56 from 54 is indicative of the difference in the number of hydrogen atoms. We note that for an ion with a given number of C and N atoms, the yield curves (and also their peaks) tend to shift to higher electron energies as the number of H atoms decreases. This seems to be the case in almost all the ions we observed.

Next we consider the ions in the range of $m/z = 65$ to 70. In addition to 65 and 70, we observe 66, 67, 67.5, and 68. The ion of $m/z = 67.5$ corresponds to the doubly charged parent ion. Due to insufficient mass resolution, we are unable to separate it from $m/z = 67$ and 68 to obtain accurate numbers. Our analysis shows that at 100 eV, $m/z = 68$ has very small intensity, and the intensity of $m/z = 67.5$ is about a factor of three larger in comparison to that of $m/z = 68$. Thus, the cross section curve identified as $m/z = 67$–68 may be treated to be mainly that of the doubly charged adenine ion. This may explain the difference in the shape of the curve in comparison to others in the panel. $C_3N_2H^+$ is the composition for $m/z = 65$, while $m/z = 70$ is most likely to be $C_2N_3H_4^+$. $m/z = 66$ has been identified to be $C_3N_2H_2^+$, though, in principle, it could also be $C_2N_3^+$. The yield curves for $m/z = 65$ and $m/z = 70$ are clearly different, consistent with their composition and based on what we have seen so far from other groups of ions. The question is, based on the empirical relation we have seen so far, whether we can identify the composition of $m/z = 66$. We note that the cross section for $m/z = 66$ peaks at about 80 eV, while that of $m/z = 65$ ($C_3N_2H^+$) peaks at about 95 eV, and for $m/z = 70$ ($C_2N_3H_4^+$), it peaks at 90 eV. If $m/z = 66$ were to be $C_2N_3^+$, its yield curve should have peaked at an energy higher than 90 eV. On the other hand, if its composition is $C_3N_2H_2^+$, its yield curve should peak at an energy smaller than that of $m/z = 65$ ($C_3N_2H^+$), which it does. So one may conclude that the composition of $m/z = 66$ is $C_3N_2H_2^+$ and not $C_2N_3^+$. While it may be already known or there may be other means to identify the composition of $m/z = 66$ from adenine, what we want to show is the consistency in the relation between the composition and the shape of the ion yield curves that we have discussed so far.

In the range of $m/z = 80$ to 85, we observe three ions, 80, 81 and 82, with 81 ($C_3N_3H_3^+$) being the dominant ion. The yield curve of $m/z = 80$ is considerably different from the other two. In addition, it has much higher AE (17.5 eV in comparison to 13.5 eV for 81). The mass 80 ion could have two possible compositions. It could be $C_3N_3H_2^+$, that is, one H atom less than that of mass 81 ($C_3N_3H_3^+$), or there could be a difference in the number of C and N atoms along with appropriate number of H atoms, that is, $C_4N_2H_4^+$. Based on the difference in shape of the ion yield curve and the large difference in AE, we may argue that the composition of $m/z = 80$ is $C_4N_2H_4^+$ and not $C_3N_3H_2^+$.

The cross sections for the ions of $m/z = 91$ to 94 are relatively low, and their yield curves look almost similar, except for a slight variation in the peak positions. The compositions of the ions are most likely to be $C_4N_3H_n^+$, with n varying from 1 to 4. With only a change in the number of H atoms, it is not surprising to see them having almost similar ion yield curves. The shift in the energy of the peak cross section seems to decrease with an increase in the number of H atoms, consistent with our discussion so far.

We present cross sections for ions of $m/z = 108$ ($C_4N_4H_4^+$) and 135 (the parent ion, $C_5N_5H_5^+$) in the last panel in Figure 4. As is expected, the ion yield curves for the two look different from each other. The AE of 12.3 ± 0.5 eV for $m/z = 108$ is smaller than all other fragment ions. However, we note that this is slightly larger than other reported measurements [14–16]. The peak cross section for m/z = 108 is at 115 eV. The AE for the parent ion is 8.8 ± 0.3 eV and is consistent with previous reports. The maximum of the cross section is found to be at about 75 eV. The most notable feature in the case of the parent ion is the presence of some structure in the ion yield curve at about 35 eV.

So far, we have discussed the cross sections in terms of the AEs and the energy corresponding to the peak cross section for given sets of fragment ions. We observed that the lighter the ion, the more the tendency to increase both the AE and the peak energy. In addition, given a fixed number of C and N atoms, the addition of H atoms seems to lower these energies. The general tendency for the lighter ions to have larger peak energy was seen in the case of uracil [43]. This was explained based on the availability of states in the ionization continuum, which increases with electron energy. The shift of the peaks to higher values for lighter ions is due to the fact that these ions have more available channels of formation as compared to the heavier ones. One way this happens is through the opening of multiple ionizations and excitations of heavier ions, which fragments, giving rise to lighter ions.

Figure 5 (reproduced from [40]) shows the total ionization cross sections from threshold to 500 eV. Though already reported, we provide it here for continuity and as a marker of accuracy for the partial cross sections. As mentioned earlier, the absolute total cross sections were obtained by summing the absolute partial ion cross sections that we measured. Thus, total ion cross sections would indicate the accuracy of the measured partial ion cross sections. In the figure, our total ion cross sections are compared with the available experimental [11,16] and theoretical results [18–24]. So far, only Minaev et al. [11] have made absolute cross section measurements for adenine up to 200 eV. Their results are considerably larger than the present results and all the theoretical results. As mentioned earlier, van der Burgt et al. [16] measured the relative cross sections up to 100 eV and normalized them to absolute values using the average value of various theoretical results at 70 eV. All theoretically calculated cross section curves exhibit the typical shape, with a maximum energy between 80–90 eV and a gradual decline towards higher energies. The cross sections in our measurements peak at 100 eV. On the whole, there is reasonable agreement between our data and most of the theoretical results. However, it appears that the improved binary-encounter dipole (iBED) model employed by Huo et al. [22] provides the best agreement. We also note that the recent calculations by Tan et al. [25] (not shown in the figure) have better agreement with the present results below 100 eV. Very recently, Mendez et al. [58] provided a scaling rule for the ionization of biological molecules by fully stripped ions. Their scaling law for H^+ projectiles and our results for electron projectiles show excellent agreement for adenine along with other DNA bases. This is important in the context of the limited electron impact data with which we could compare the present results. The consistency shown in the scaling behavior indirectly certifies the accuracy of the present total ion cross sections and, consequently, the present partial ion cross sections.

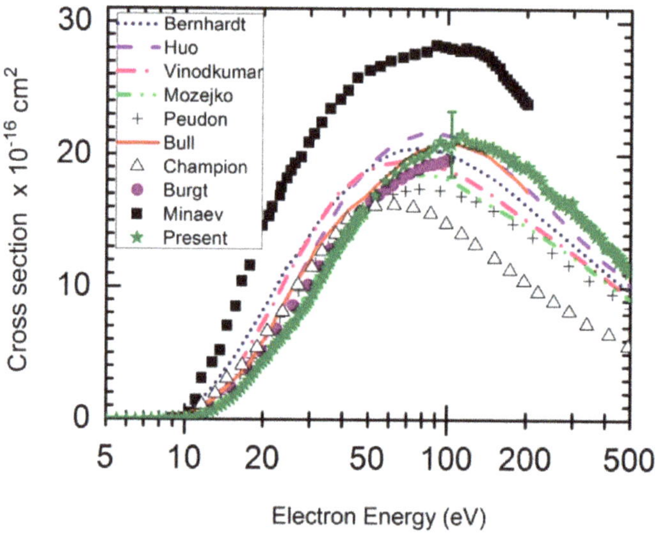

Figure 5. Total ionization cross section for the adenine molecule and its comparison with the existing data (Reproduced from Ref. [40]).

4. Conclusions

In this work, we used a crossed electron-molecular beam experiment along with the Relative Flow Technique (RFT) to measure the absolute partial ionization cross sections of adenine molecules, which exist in solid form at room temperature. The total ionization cross sections obtained from summing the partial cross sections are compared with the theoretical calculations and are found in reasonable agreement. Our measured cross section values are in close agreement with the theoretical calculation by Huo et al. [22] using the improved binary-encounter dipole (iBED) model. The relative ion intensities (mass spectrum) and appearance energies measured are also found in reasonable agreement with the existing data. The most abundant fragment cations from adenine include $C_nH_nN_n^+$ (n = 5, 4, 3, 2, 1) at m/z of 135 ($C_5N_5H_5$+), 108 ($C_4N_4H_4^+$), 81 ($C_3N_3H_3^+$), 54 ($C_2N_2H_2^+$), 27 (CNH^+), and $HCNH^+$ as determined from the experiment. The dominance of $C_nH_nN_n^+$ (n = 1 to 5) confirmed by our cross section measurements shows that the most thermodynamically favored pathway for adenine dissociation due to electron ionization appears to be the loss of HCN molecules in succession. This supports the idea of formation of adenine by successive addition of HCN units and may help us in understanding the formation of adenine in space.

Comparison of the ionization yield curves for various sets of fragment ions provided some interesting, though not unexpected results. We find that lighter ions tend to have higher appearance energies as well as higher energies at which the cross section peaks. Within a group of ions with a given number of C and N atoms, we find that there is a systematic shift in the abovementioned energies as a function of the number of H atoms. We also find that we are able to use this information to identify the composition of a given ion where more than one possibility exists. The fact that adenine is a relatively simple system with only three species of atoms was helpful in this respect. It would be interesting to investigate this idea in other similar molecules. Rich possibilities exist for the further investigation of dissociative ionization of adenine using momentum imaging. Finally, we hope that the measured partial cross sections will be immediately useful in biochemical modelling and Monte Carlo track simulations to understand the damage mechanism in the living cells and in other applications.

Author Contributions: Conceptualization of the experiment: E.K. Method and measurements: carried out by M.A.R. under the guidance of E.K. Analysis of the data and preparation of the manuscript: all authors. All authors have read and agreed to the published version of the manuscript.

Funding: This research received no external funding.

Institutional Review Board Statement: Not applicable.

Informed Consent Statement: Not applicable.

Data Availability Statement: Data are contained within the article.

Conflicts of Interest: The authors declare no conflict of interest.

References

1. Ward, J.F.; Webb, C.F.; Limoli, C.L.; Milligan, J.R. *Ionizing Radiation Damage to DNA: Molecular Aspects*; Wiley-Liss: New York, NY, USA, 1990; p. 43.
2. Huels, M.A.; Hahndorf, I.; Illenberger, E.; Sanche, L. Resonant dissociation of DNA bases by subionization electrons. *J. Chem. Phys.* **1998**, *108*, 1309–1312. [CrossRef]
3. Boudaöffa, B.; Cloutier, P.; Hunting, D.; Huels, M.A.; Sanche, L. Resonant Formation of DNA Strand Breaks by Low-Energy (3 to 20 eV) Electrons. *Science* **2000**, *287*, 1658–1660. [CrossRef] [PubMed]
4. Hunniford, C.A.; McCullough, R.W.; Jeremy, R.; Davies, H.; Timson, D. DNA damage by low-energy ions. *Biochem. Soc. Trans.* **2009**, *37*, 893–896. [CrossRef]
5. Baccarelli, I.; Bald, I.; Gianturco, F.A.; Illenberger, E.; Kopyra, J. Electron-induced damage of DNA and its components: Experiments and theoretical models. *Phys. Rep.* **2011**, *508*, 1–44. [CrossRef]
6. Nikjoo, H.; Neill, P.O.; Terrissol, M.; Goodhead, D.T. Quantitative modelling of DNA damage using Monte Carlo track structure method. *Radiat. Environ. Biophys.* **1999**, *38*, 31–38. [CrossRef] [PubMed]
7. Rabus, H.; Nettelbeck, H. Nanodosimetry: Bridging the Gap to Radiation Biophysics. *Radiat. Meas.* **2011**, *46*, 1522–1528. [CrossRef]
8. Nettelbeck, H.; Rabus, H. Nanodosimetry: The missing link between radiobiology and radiation physics? *Radiat. Meas.* **2011**, *46*, 893–897. [CrossRef]
9. Schardt, D.; Elsasser, T.; Schulz-Ertner, D. Heavy-ion tumor therapy: Physical and radiobiological benefits. *Rev. Mod. Phys.* **2010**, *82*, 383. [CrossRef]
10. Shafranyosh, I.I.; Sukhoviya, M.I. Inelastic collisions of the uracil molecules with electrons. *J. Chem. Phys.* **2012**, *137*, 184303–184306. [CrossRef]
11. Minaev, B.F.; Shafranyosh, M.I.; Svida, Y.Y.; Sukhoviya, M.I.; Shafranyosh, I.I.; Baryshnikov, G.V.; Minaev, V.A. Fragmentation of the adenine and guanine molecules induced by electron collisions. *J. Chem. Phys.* **2014**, *140*, 175101. [CrossRef] [PubMed]
12. Shafranyosh, I.I.; Sukhoviya, M.I.; Shafranyosh, M.I. Absolute cross sections of positive- and negative-ion production in electron collision with cytosine molecules. *J. Phys. B.* **2006**, *39*, 4155–4162. [CrossRef]
13. Shafranyosh, I.I.; Sukhoviya, M.I.; Shafranyosh, M.I.; Shimon, L.L. Formation of positive and negative ions of thymine. *Tech. Phys.* **2008**, *53*, 1536–1540. [CrossRef]
14. Jochims, H.W.; Schwell, M.; Baumgartel, H.; Leach, S. Photoion mass spectrometry of adenine, thymine and uracil in the 6–22 eV photon energy range. *Chem. Phys.* **2005**, *314*, 263–282. [CrossRef]
15. Dawley, M.M.; Tanzer, K.; Cantrell, W.A.; Plattner, P.; Brinkmann, N.R.; Scheier, P.; Denifl, S.; Ptasinska, S. Electron ionization of the nucleobases adenine and hypoxanthine near the threshold: A combined experimental and theoretical study. *Phys. Chem. Chem. Phys.* **2014**, *16*, 25039–25053. [CrossRef] [PubMed]
16. van der Burgt, P.J.M.; Finnegan, S.; Eden, S. Electron impact fragmentation of adenine: Partial ionization cross sections for positive fragments. *Eur. Phys. J. D* **2015**, *69*, 173. [CrossRef]
17. Ostroverkh, A.; Zavilopulo, A.; Shpenik, O. Ionization of guanine, adenine and thymine molecules by electron impact. *Eur. Phys. J. D* **2019**, *73*, 38. [CrossRef]
18. Bernhardt, P.H.; Paretzke, H.G. Calculation of electron impact ionization cross sections of DNA using the Deutsch–Märk and Binary–Encounter–Bethe formalisms. *Int. J. Mass Spectrom.* **2003**, *223–224*, 599–611. [CrossRef]
19. Możejko, P.; Sanche, L. Cross section calculations for electron scattering from DNA and RNA bases. *Radiat. Environ. Biophys.* **2003**, *42*, 201–211. [CrossRef] [PubMed]
20. Peudon, A.; Edel, S.; Terrissol, M. Molecular basic data calculation for radiation transport in chromatin. *Radiat. Prot. Dosim.* **2006**, *122*, 128–135. [CrossRef]
21. Bull, J.N.; Lee, J.W.L.; Vallance, C. Absolute electron total ionization cross-sections: Molecular analogues of DNA and RNA nucleobase and sugar constituents. *Phys. Chem. Chem. Phys.* **2014**, *16*, 10743–10752. [CrossRef] [PubMed]
22. Huo, W.M.; Dateo, C.E.; Fletcher, G.D. Molecular data for a biochemical model of DNA damage: Electron impact ionization and dissociative ionization cross sections of DNA bases and sugar-phosphate backbone. *Radiat. Meas.* **2006**, *41*, 1202–1208. [CrossRef]
23. Vinodkumar, M.; Limbachiya, C.; Barot, M.; Swadia, M.; Barot, A. Electron impact total ionization cross sections for all the components of DNA and RNA molecule. *Int. J. Mass Spectrom.* **2013**, *339–340*, 16–23. [CrossRef]

24. Champion, C. Quantum-mechanical predictions of electron-induced ionization cross sections of DNA components. *J. Chem. Phys.* **2013**, *138*, 184306. [CrossRef] [PubMed]
25. Tan, H.Q.; Mi, Z.; Bettiol, A.A. Simple and universal model for electron-impact ionization of complex biomolecules. *Phys. Rev. E* **2018**, *97*, 032403. [CrossRef]
26. Bauer, C.A.; Grimme, S. Elucidation of electron ionization induced fragmentations of adenine by semiempirical and density functional molecular dynamics. *J. Phys. Chem. A* **2018**, *118*, 11479. [CrossRef] [PubMed]
27. Chakraborti, S.; Chakraborti, S.K. Can DNA bases be produced during molecular cloud collapse? *Astron. Astrophys.* **2000**, *354*, L6–L8.
28. Saladino, R.; Crestini, C.; Costanzo, G.; Negri, R.; di Mauro, E. A possible prebiotic synthesis of purine, adenine, cytosine, and 4(3H)-pyrimidinone from formamide: Implications for the origin of life. *Bioorg. Med. Chem.* **2001**, *9*, 1249–1253. [CrossRef]
29. Gupta, V.P.; Tandon, P.; Rawat, P.; Singh, R.N.; Singh, A. Quantum chemical study of a new reaction pathway for the adenine formation in the interstellar space. *Astron. Astrophys.* **2011**, *528*, A129. [CrossRef]
30. Evans, N.L.; Ullrich, S.; Bennett, C.J.; Kaiser, R.I. On the interaction of adenine with ionizing radiation: Mechanistical studies and astrobiological implications. *Astrophys. J.* **2011**, *730*, 69. [CrossRef]
31. Callahan, M.P.; Smith, K.E.; Cleaves, H.J.; Ruzicka, J.; Stern, J.C.; Glavin, D.P.; House, C.H.; Dworkin, J.P. Carbonaceous meteorites contain a wide range of extraterrestrial nucleobases. *Proc. Natl. Acad. Sci. USA* **2011**, *108*, 13995. [CrossRef]
32. Martins, Z.; Botta, O.; Fogel, M.L.; Sephton, M.A.; Glavin, D.P.; Watson, J.S.; Dworkin, J.P.; Schwartz, A.W.; Ehrenfreund, P. Extraterrestrial nucleobases in the Murchison meteorite. *Earth Planet. Sci. Lett.* **2008**, *270*, 130–136. [CrossRef]
33. Zaleski, D.P.; Seifert, N.A.; Steber, A.L.; Muckle, M.T.; Loomis, R.A.; Corby, J.F.; Martinez, O.; Crabtree, K.N.; Jewell, P.R.; Hollis, J.M.; et al. Detection of E-cyanomethanimine toward Sagittarius B2(N) in the Green Bank telescope primos survey. *Astrophys. J. Lett.* **2013**, *765*, L10. [CrossRef]
34. Stoks, P.G.; Schwartz, A.W. Uracil in carbonaceous meteorites. *Nature* **1979**, *282*, 709–710. [CrossRef]
35. Joyce, G.F. The antiquity of RNA-based evolution. *Nature* **2002**, *418*, 214–221. [CrossRef]
36. Ziurys, L.M. The chemistry in circumstellar envelopes of evolved stars: Following the origin of the elements to the origin of life. *Proc. Natl. Acad. Sci. USA* **2006**, *103*, 12274–12279. [CrossRef] [PubMed]
37. Oro, J.; Kimball, A.P. Synthesis of purines under possible primitive earth conditions: II. Purine intermediates from hydrogen cyanide. *Arch. Biochem. Biophys.* **1962**, *96*, 293–313. [CrossRef]
38. Sanchez, R.A.; Ferris, J.P.; Orgel, L.E. Studies in prebiotic synthesis. IV. Conversion of 4-aminoimidazole-5-carbonitrile derivatives to purines. *J. Mol. Biol.* **1968**, *38*, 121–128. [CrossRef]
39. Volkenshtein, M.V. *Biophysics*; Mir Publishers: Moscow, Russia, 1983.
40. Rahman, M.A.; Krishnakumar, E. Communication: Electron ionization of DNA bases. *J. Chem. Phys.* **2016**, *144*, 161102. [CrossRef]
41. Srivastava, S.K.; Chutjian, A.; Trajmar, S. Absolute elastic differential electron scattering cross sections in the intermediate energy region. I. H_2. *J. Chem. Phys.* **1975**, *63*, 2659–2665. [CrossRef]
42. Rahman, M.A.; Gangopadhyay, S.; Limbachiya, C.; Joshipura, K.N.; Krishnakumar, E. Electron ionization of NF_3. *Int. J. Mass Spectrom.* **2012**, *319–320*, 48–54. [CrossRef]
43. Rahman, M.A.; Krisnakumar, E. Absolute partial and total electron ionization cross sections of uracil. *Int. J. Mass Spectrom.* **2015**, *392*, 145. [CrossRef]
44. Zielenkiewicz, W.J. Enthalpies of Sublimation and Vapor Pressures of Adenine, 1-Methyladenine, 2-Methyladenine, 3-Methyladenine, and 8-Methyladenine. *Chem. Eng. Data* **2000**, *45*, 626. [CrossRef]
45. Rejoub, R.; Lindsay, B.G.; Stebbings, R.F. Determination of the absolute partial and total cross sections for electron-impact ionization of the rare gases. *Phys. Rev. A* **2002**, *65*, 042713. [CrossRef]
46. Krishnakumar, E.; Srivastava, S.K. Ionisation cross sections of rare-gas atoms by electron impact. *J. Phys. B At. Mol. Opt. Phys* **1988**, *21*, 1055–1082. [CrossRef]
47. Straub, H.C.; Renault, P.; Lindsay, B.G.; Smith, K.A.; Stebbings, R.F. Absolute partial and total cross sections for electron-impact ionization of argon from threshold to 1000 eV. *Phys. Rev. A* **1995**, *52*, 1115–1124. [CrossRef]
48. Ma, C.; Sporleder, C.R.; Bonham, R.A. A pulsed electron beam time of flight apparatus for measuring absolute electron impact ionization and dissociative ionization cross sections. *Rev. Sci. Instrum.* **1991**, *62*, 909–924. [CrossRef]
49. Rice, J.M.; Dudek, G.O. Mass spectra of nucleic acid derivatives. **II.** Guanine, adenine, and related compounds. *J. Am. Chem. Soc.* **1967**, *89*, 2719–2725. [CrossRef]
50. Data Compiled by: NIST Mass Spectrometry Data Center, Wallace, W.E. Adenine Mass Spectrum (Electron Ionization). Available online: https://webbook.nist.gov/cgi/cbook.cgi?ID=C73245&Units=SI&Mask=200#Mass-Spec (accessed on 1 September 2015).
51. Krishnakumar, E.; Srivastava, S.K. Cross sections for the production of N_2^+, N^++N_2^{2+} and N^{2+} by electron impact on N_2. *J. Phys. B At. Mol. Opt. Phys.* **1990**, *23*, 1893–1903. [CrossRef]
52. Mark, T.D. Cross section for single and double ionization of N_2 and O_2 molecules by electron impact from threshold up to 170 eV. *J. Chem. Phys.* **1975**, *63*, 3731–3736. [CrossRef]
53. Plutzer, C.; Kleinermanns, K. Tautomers and electronic states of jet-cooled adenine investigated by double resonance spectroscopy. *Phys. Chem. Chem. Phys.* **2002**, *4*, 4877–4882. [CrossRef]
54. Pilling, S.; Lago, A.F.; Coutinho, L.H.; de Castilho, R.B.; de Souza, G.G.B.; de Brito, A.N. Dissociative photoionization of adenine following valence excitation. *Rapid Commun. Mass Spectrom.* **2007**, *21*, 3646–3652. [CrossRef] [PubMed]

55. Sethi, S.K.; Gupta, S.P.; Jenkins, E.E.; Whitehead, C.W.; Townsend, L.B.; McCloskey, J.A. Mass spectrometry of nucleic acid constituents. Electron ionization spectra of selectively labeled adenines. *J. Am. Chem. Soc.* **1982**, *104*, 3349–3353. [CrossRef]
56. Porter, Q.N.; Baldas, J. *Mass Spectrometry of Heterocyclic Compounds*; Wiley-Interscience: New York, NY, USA, 1971.
57. Goto, T.; Tatematsu, A.; Matsuura, S. Organic mass spectrometry. I. Mass spectra of pteridine, methylpteridines, and hydroxypteridines. *J. Org. Chem.* **1965**, *30*, 1844–1846. [CrossRef]
58. Mendez, A.M.P.; Montanari, C.C.; Miraglia, J.E. Scaling rules for the ionization of biological molecules by highly charged ions. *J. Phys. B At. Mol. Opt. Phys.* **2020**, *53*, 175202. [CrossRef]

atoms

MDPI

Article

Dynamics of Site Selectivity in Dissociative Electron Attachment in Aromatic Molecules

Vishvesh Tadsare, Sukanta Das, Samata Gokhale, E. Krishnakumar and Vaibhav S. Prabhudesai *

Tata Institute of Fundamental Research, Colaba, Mumbai 400005, India
* Correspondence: vaibhav@tifr.res.in

Abstract: Dissociative electron attachment has shown site selectivity in aliphatic molecules based on the functional groups present in them. This selectivity arises from the core excited resonances that have excited parent states localized to a specific site of the functional group. Here, we show that such site selectivity is also observed in the amine group when present in aromatic molecules. However, the proximity of the aromatic ring to the functional group under investigation has a substantial effect on the dissociation dynamics. This effect is evident in the momentum distribution of the hydride ions generated from the amine group. Our results unravel the hitherto unknown facets of the site selectivity in aromatic organic molecules.

Keywords: electrons; molecules; cross-sections; dynamics

Citation: Tadsare, V.; Das, S.; Gokhale, S.; Krishnakumar, E.; Prabhudesai, V.S. Dynamics of Site Selectivity in Dissociative Electron Attachment in Aromatic Molecules. *Atoms* **2022**, *10*, 98. https://doi.org/10.3390/atoms10040098

Academic Editors: Rajesh Srivastava and Dmitry V. Fursa

Received: 1 September 2022
Accepted: 17 September 2022
Published: 22 September 2022

1. Introduction

Low-energy electron-induced chemistry plays a vital role in many phenomena, ranging from radiation biology to astrochemistry. Dissociative electron attachment (DEA) is a dominant channel in these processes where the electron energy is intrinsically translated to the nuclear motion via the dynamics of the negative ion resonance state (NIRS). The most fascinating aspect of DEA is the site selectivity shown in the dissociation process, as manifested in the hydride ion formation [1]. This site selectivity directly correlates with the functional group present at the site. Hence, it has been seen even in larger molecules such as DNA bases [2]. This site selectivity is beyond the conventional threshold energy-based selectivity shown in the dissociation processes [3]. The NIRSs that participate in the electron attachment and give site selectivity are core excited resonances. The underlying parent excited neutral states for these NIRSs show the localization of the excited molecular orbital [4]. The excitation of non-bonding electrons from a specific site to the anti-bonding molecular orbital with the capture of the incoming electron to the same or other orbital results in the localization of the dissociation process. For example, hydride ion formation in ammonia peaks at 5.5 eV electron energy. This peak arises from a core excited Feshbach resonance. The underlying parent neutral excited state is the first triplet excited state of ammonia known to dissociate along with the $H-NH_2$ bond [5]. This excited state is formed by the excitation of the lone pair of electrons from the N atom to the Rydberg type orbital $3sa_1'$. This localized excited state provides the site selectivity to the DEA process. The localization due to the excitation of the lone pair of electrons in ammonia remains almost unaffected in simple saturated aliphatic amines such as n-propyl amine [4,6]. However, as in the unsaturated amines and particularly in the aromatic amines, the lone pair of electrons from the N-site may get easily influenced by the delocalized π electron cloud of the phenyl group, it would be interesting to see its effect on the site selectivity in the DEA process.

In DEA to various amines, the systematic difference between the saturated and unsaturated amines has been seen in terms of the ion yields and their peaks [7], which indicates the presence of low-lying shape resonance at lower electron energy in the $(M-H)^-$ channel around 2 eV. The 5.5 eV peak in the H^- channel, which has been seen in ammonia and n-propyl amine (at 5.7 eV), is also consistently seen in the $(M-H)^-$ channel in the saturated amines with varying energy positions [7]. However, this peak loses its intensity substantially in the aromatic or unsaturated amines, and the low-energy shape resonance dominates the $(M-H)^-$ DEA channel. The presence of this resonance is understood to be due to π and π^* orbitals in the unsaturated amines, which influence the Rydberg-type orbitals. With this observation in mind, we have investigated the DEA dynamics of the H^- channel in the aromatic molecules aniline and benzylamine using the velocity slice imaging technique.

In aniline, the amine group is directly attached to the phenyl group, making it an unsaturated amine, whereas, in benzylamine, the amine group is attached to the benzyl group, making it another saturated amine. However, the benzylamine also contains the aromatic ring, giving a test bed to explore its influence on the site selectivity of DEA and its underlying dynamics.

The aromatic amines are also important from a biological point of view. Aromatic amines are present in many biological molecules, including DNA bases adenine, guanine, and cytosine, and, as mentioned earlier, it is well-recognized that the reactions induced by low-energy electrons in living cells represent an important step toward radiation damage [8,9]. Understanding the DEA dynamics in these molecules will also serve as a stepping stone to studying the DEA in higher aromatic compounds and complex biomolecules. We also present the absolute cross-sections for the DEA process in these two molecules as it is closely related to the dynamics of the process and is of importance in theoretical modeling as well as for possible use in practical applications.

2. Experimental Setup

We measured the kinetic energy and angular distribution of hydride anions formed in DEA to aniline and benzylamine by the velocity slice imaging (VSI) technique. Although the experimental setup for these measurements is described earlier, here we give its brief description [10]. We performed the experiments by crossing the magnetically collimated pulsed electron beam at right angles with the effusive molecular beam generated using a capillary array. The molecular beam coincided with the axis of the VSI spectrometer. The electron current was measured using a Faraday cup placed against the electron gun across the interaction region of the spectrometer. The pulsed electrostatic field on the pusher electrode extracted the ions. A four-element electrostatic lens with the appropriate voltage condition velocity focused the ions onto the position-sensitive detector made of a pair of microchannel plates (MCPs) of 75 mm in diameter in the chevron geometry, followed by a phosphor screen. We used a CCD camera to capture the illuminated spots on the phosphor screen due to the ion hit on the detector. While determining the ion yield as a function of the incoming electron energy, we used the spectrometer in the mass-spectrometer mode. In this mode, we applied DC electric potentials to the MCPs and the phosphor screen. We counted the ions by detecting the electrical signal pulses on the MCP back electrode. The voltage pulses were amplified and fed to a multiple-event time digitizer to obtain the time-of-flight mass spectra. Using these mass spectra as a function of electron energy, we determined the relative cross-sections for different DEA channels. We used the relative flow technique to put these cross-sections on the absolute scale using the mass spectra obtained for O^- from molecular oxygen. To obtain accurate absolute cross-sections, we ensured the complete collection of all the ions onto the detector (as confirmed by the VSI data) and the elimination of the bias voltage dependence on the detection efficiency as a function of mass-to-charge ratios. We used the arrival time of ions of given m/e to determine the detector pulsing delay in the imaging mode. In this mode, we pulsed the MCP back voltage at an appropriate time delay to capture the signal from the ions of interest. We recorded the images on the CCD camera and analyzed the ion hit distribution in the offline mode using a homebuilt Matlab-based data analysis program. We used 80 ns pulses on the detector to obtain the slice images.

For creating the molecular beam, we used aniline (purity > 99.5%) and benzylamine (purity > 99.5%) from Sigma-Aldrich. The liquid sample was stored in an evacuated glass bulb connected to the vacuum chamber via a glass-to-metal seal using a $\frac{1}{4}$-inch stainless steel tube through a leak valve (make: Granville Phillips). To eliminate any dissolved gases, including water vapor, the sample liquid in the bulb was pumped by a rotary pump until its volume was reduced to 1/3rd before using it as the target source. The sample vapor beam was introduced in the chamber in two ways: (i) through the capillary array (we term this as a crossed-beam measurement) and (ii) directly into the chamber (we term this as a static gas background measurement). The overall chamber pressure rises when the target vapor is flown through the capillary array. As the electron beam passes through this raised background gas, it contributes to the measured mass spectrum. The static gas background measurements are carried out to subtract these background counts from the crossed beam measurements. The two spectra are subtracted on proper current and pressure normalization. The details of this technique to measure the absolute cross-section are given in ref. [4].

3. Results and Discussions

3.1. Absolute Cross-Sections

The earlier measurements of the negative ion formation in aniline by resonant electron capture by Pikhtovnikov et al. show $(M-H)^-$ ion signal peaking at 2.6, 5.3, 6.3, and 8.3 eV; $(M-2H)^-$ ion signal peaking at 5.3, 7, 9.2, and 10.6 eV; $(M-NH_2)^-$ ion signal peaking at 6.7 eV; CN^- peaking at 6 eV; C_2H^- peaking at 9.7 eV; NH_2^- peaking at 7.1 eV; and NH^- ion signal peaking at 9.2 eV [11]. They found the $(M-H)^-$ at 8.3 eV, NH_2^-, and CN^- as the most dominant channels. However, DEA to benzylamine has not been reported in the literature so far.

We found H^- and CN^- as the most dominant channels in our measurements of DEA to aniline and benzylamine. The absolute cross-sections as a function of electron energy for these channels from both the molecules are shown in Figure 1.

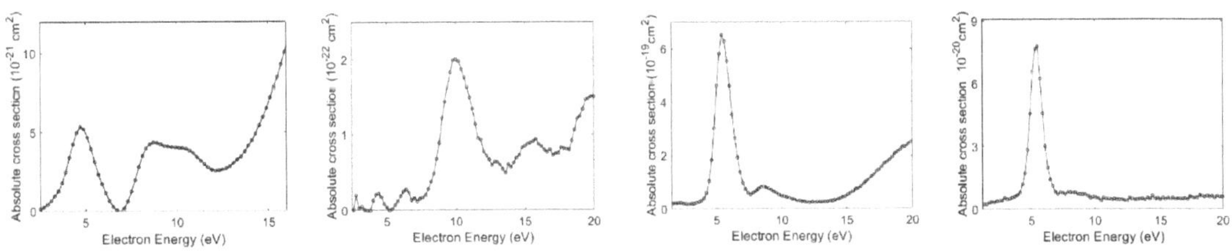

(a) (b) (c) (d)

Figure 1. DEA absolute cross-sections measured for (**a**) H^- and (**b**) CN^- channels from aniline and (**c**) H^- and (**d**) CN^- channels from benzylamine as a function of electron energy.

As the figure shows, both the molecules show H^- as the most dominant channel. However, the cross-section for this channel from aniline is two orders of magnitude lower than that for benzylamine. This observation is consistent with the previous measurements of the $(M-H)^-$ channel from the saturated and unsaturated amines [7]. Interestingly, the H^- channel also shows a similar trend. The H^- cross-section shows three peaks at 5, 8.7, and 9.9 eV, whereas in the benzylamine, the H^- cross-section peaks at 5.5 and 8.4 eV. The CN^- channel peaks at 10 and 5.5 eV, respectively. For benzylamine, the absolute cross-section for NH_2^- also peaks at 5.5 eV, whereas this channel is too weak in aniline. The measured absolute cross-section values for the DEA channels from both the molecules are given in Table 1.

Table 1. Absolute cross-sections for the dominant channels observed in DEA to aniline and benzylamine. The absolute cross-sections for ammonia and n-propylamine are also given. In all these values, the typical uncertainty is about 15%.

	H^-		NH_2^-		CN^-	
Molecule	**Peak Position (eV)**	**Cross-Section (cm^2)**	**Peak Position (eV)**	**Cross-Section (cm^2)**	**Peak Position (eV)**	**Cross-Section (cm^2)**
Aniline	5.0 8.7 9.9	5.3×10^{-21} 4.3×10^{-21} 4.3×10^{-21}			10.0 15.4	2.0×10^{-22} 1.0×10^{-22}
Benzylamine	5.5 8.4	6.5×10^{-19} 9.0×10^{-20}	5.5	6.0×10^{-20}	5.5 8.3	8×10^{-20} 8×10^{-21}
Ammonia [1]	5.7 10.5	2.3×10^{-18} 5.0×10^{-19}	5.9 10.2	1.6×10^{-18} 9.0×10^{-20}		
n-propylamine [2]	5.2 8.8	5.2×10^{-20} 1.7×10^{-20}				

[1] Experimental value from reference [12], [2] Experimental value from reference [4].

The absolute cross-section values obtained for the benzylamine and aniline follow the trend observed in an earlier work on the saturated and unsaturated amines [7]. We have determined the origin of the hydride ions at various observed peaks by measuring the ion yield curves for the H^- and D^- ions from partially deuterated aniline ($C_6D_5NH_2$). The results are shown in Figure 2.

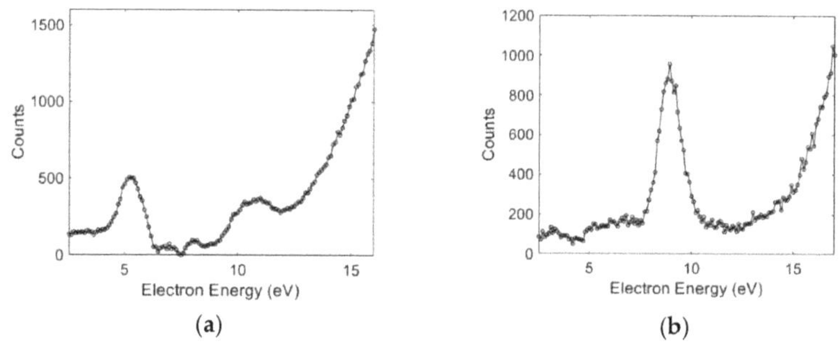

Figure 2. The ion yield curves as a function of electron energy for (**a**) H^- and (**b**) D^- ions from DEA to partially deuterated aniline ($C_6D_5NH_2$).

As shown in Figure 2a, the 5 and 9.9 eV peaks can be attributed to the N − H site in the molecule. The 8.7 eV peak can be attributed to the phenyl part of the molecule (Figure 2b). This clearly shows the site selectivity observed in the aromatic amine. Interestingly, the presence of the phenyl group does not substantially alter the DEA peak from the N − H site compared to that from ammonia. With the observed analogy with the amine group in the aliphatic compounds, we attribute the 5.5 eV peak in benzylamine to the amine site and that at the 8.4 eV to the phenyl group.

Photodissociation measurements by King et al. [13] on aniline revealed that the second excited singlet electronic state of aniline corresponds to a $\pi \rightarrow 3s$ Rydberg electronic transition, calculated to lie at 4.53 eV. They anticipated that the 3*s* Rydberg electron density in this latter excited state evolves into an σ^* character upon the extension of one of the N − H bonds, which shows a conical intersection with the ground state at the stretched N–H bond. It contributes to the direct dissociation of this bond, i.e., the second singlet excited state of aniline is expected to be essentially repulsive and of 1 $\pi\sigma^*$ character. Hence, this state may act as the parent state for the anion state of aniline, contributing to the H^- channel that peaks at 5 eV. The peak in the hydride ions from benzylamine was at 5.5 eV, which is closer to that obtained in ammonia (5.5 eV) and n-propylamine (5.7 eV).

The first ionization energy of aniline is at 8 eV, and the second is at 9.1 eV. Since the H^- comes at 8.9 eV at the second resonance, we conclude that the excitation of electrons from HOMO may not play any role in this resonance. The HOMO-1 has a contribution largely from the delocalized electrons from the phenyl ring. The 8.9 eV resonance, as observed in aniline and its deuterated form, may arise from the excitation of this HOMO-1 orbital. In an attempt to find the parent state of this resonance, we compared the DEA with the VUV absorption [14] and electron energy loss spectra [15] reported earlier. From these results, we note that the n $\rightarrow$ 3p Rydberg transition at about 8.5 eV may be the parent state of the DEA resonance at 8.7 eV. Interestingly, DEA to benzene also shows a peak in the $C_6H_5^-$ channel around 8 eV [16]. Benzylamine also shows a peak at 8.4 eV. We may conclude that the molecule's phenyl part may play a significant role in this resonance. Benzylamine also shows a very weak shoulder near 10 eV. Based on the comparison with the DEA studies on ammonia and methane, we propose the origin of this shoulder to be from the $N-H$ site, like in ammonia, or from the $C-H$ site of CH_2, like in methane.

To probe the effect of the delocalized electrons on the site selectivity of DEA and the underlying dynamics, we performed the angular and kinetic energy distribution measurements of the ions formed by DEA to both the molecules using the VSI technique.

3.2. Angular and Kinetic Energy Distributions

We measured the VSI images of the H^- and CN^- ion momentum distribution at various peaks observed in the cross-section measurements of both the molecules. We discuss the details of the kinetic energy and angular distributions obtained from these measurements below.

H^- Ions

The VSIs obtained for the H^- ions near the first peak, i.e., at 4.7 eV in aniline and 5.5 eV in benzylamine, are shown in Figure 3, along with the angular distributions obtained from the images. The VSI obtained for H^- ions from benzylamine shows considerable similarity with that obtained from ammonia [17] and from n-propylamine [6]. However, the image obtained for aniline shows some difference. The corresponding angular distributions are shown in Figure 3c,d. As discussed earlier, the dynamics described for the photodissociation of aniline at 4.53 eV have the origin in the $\pi \rightarrow$ 3s Rydberg electronic transition, which further passes on to the σ^* orbital. It results in the cleavage of the $N-H$ bond. As the N atom is present next to the phenyl ring, the delocalized π electrons would influence its lone pair of electrons. This, it appears, is not only affecting the cross-section of the hydride ion formation channel through DEA but also the angular distribution.

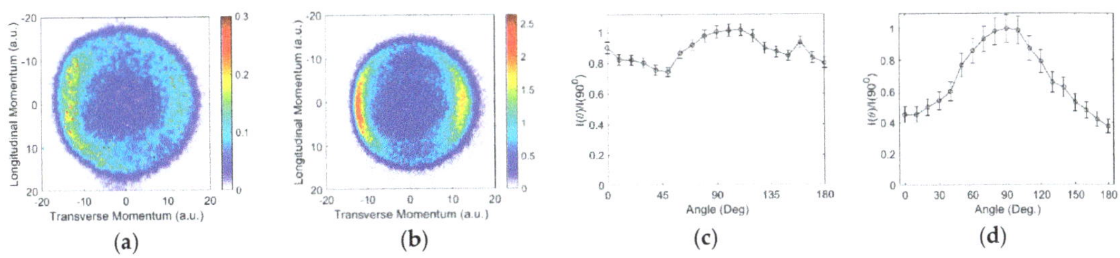

Figure 3. Momentum images obtained for the H^- ions from (**a**) aniline at 4.7 eV and (**b**) benzylamine at 5.5 eV. The incoming electron beam direction is from top to bottom; (**c**,**d**) are the respective angular distributions.

The angular distribution obtained for aniline in this channel is considerably flatter, indicating internal dynamics involved in the dissociation process. The dynamics might be in the form of changes in the bond lengths and/or angles to circumvent the barrier present in the direct dissociation path. In comparison, we find that the angular distribution for the H^- ions from benzylamine resembles that for ammonia. The cross-section is also substantially higher than aniline and compares well with n-propylamine [4]. We have also determined the kinetic energy distribution of the H^- ions, as shown in Figure 4.

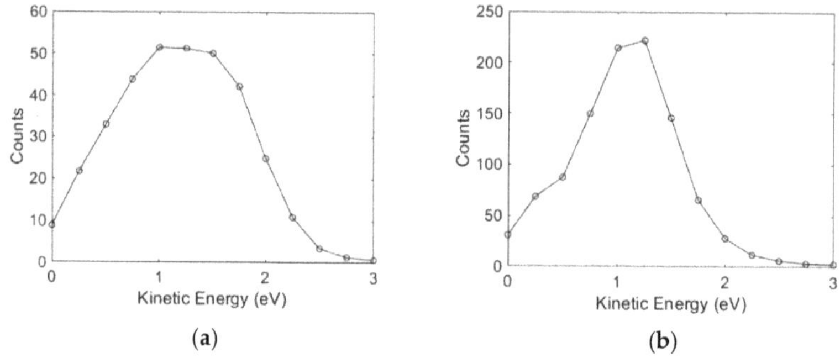

Figure 4. Kinetic energy distribution of the H^- ions obtained from (**a**) aniline at 4.7 eV and (**b**) benzylamine at 5.5 eV electron energy.

The kinetic energy distribution for H^- ions from aniline peaks at 1.25 eV and extends up to 2 eV. The heat of formation of the aniline radical (C_6H_5NH) can be estimated using its electron affinity (1.61 eV) and the heat to deprotonation reaction of the aniline molecule (1540 kJ/mol) [18]. Based on the heat of formation for H (218 kJ/mol), $C_6H_5NH_2$ (87 kJ/mol), and C_6H_5NH (252 kJ/mol) and the electron affinity of H (0.75 eV), we estimate the threshold for the H^- ion formation from the $N-H$ site via two body fragmentation as 3.2 eV. Based on the $N-H$ bond dissociation energy in aniline as 386.2 kJ/mol, the other estimate of the threshold for this channel is 3.25 eV [19]. At the 4.7 eV electron energy, the excess energy in the system would be about 1.5 eV (1.45 eV), out of which the maximum of 1.48 eV (1.43 eV) would appear as the kinetic energy of H^- ion. As most of the excess energy appears as the kinetic energy of the hydride ion, about 0.3 eV energy appears in the internal excitation of the aniline radical. However, the images obtained across the peak do not show a substantial increase in the kinetic energy of the hydride ion, indicating that the excess energy goes as the internal energy of the aniline radical. This is an important feature indicating that there might be a barrier to the direct dissociation channel, and the system undergoes dissociation with a considerable reshaping of the molecule.

For benzylamine, the bond dissociation energy of $C_6H_5CH_2HN-H$ is 418.3 kJ/mol, which makes the threshold for the H^- ion formation from the $N-H$ site 3.58 eV [20]. Hence, at 5.5 eV electron energy, the fragments would carry a maximum of 1.92 eV energy, and that would translate to the maximum of 1.9 eV as the kinetic energy of the H^- ion. The observed kinetic energy distribution at 5.5 eV peaks around 1.25 eV and extends up to 2 eV. This also clearly indicates the two-body dissociation.

The VSI and the kinetic energy distribution obtained at the second peak in the cross-section for both molecules are shown in Figure 5. The VSIs for both the molecules show a relatively big blob with a considerable intensity throughout. For both the molecules, the kinetic energy distribution obtained is spread over a similar energy range and is flatter for benzylamine than for aniline.

In the absence of the $C-H$ bond dissociation energy for the phenyl part of both compounds, we consider the upper limit for this to be the same as for benzene. The $C-H$ bond dissociation energy in benzene is 474 kJ/mol, making the threshold for this channel 4.2 eV. For benzylamine, the bond dissociation energy of the $C-H$ from the CH_2 part is 368 kJ/mol [20]. This makes the threshold for the H^- ion formation from this site 3.06 eV.

Based on these thresholds and the kinetic energy distributions observed in the H^- channel from the $C-H$ site, we conclude that the internal excitation of the molecular neutral fragment is in the range of 2.5 to 4.5 eV, which is sufficient to make this radical break further. The blob observed in the VSI and the overall energetics point to a possible many-body break-up. Moreover, considering the amount of excess energy in the internal excitation, we cannot rule out the ring-breaking dynamics as well. This conclusion is in accordance with the aliphatic compounds where the C − H bond breaks have been associated with the many-body break-up mechanism [21].

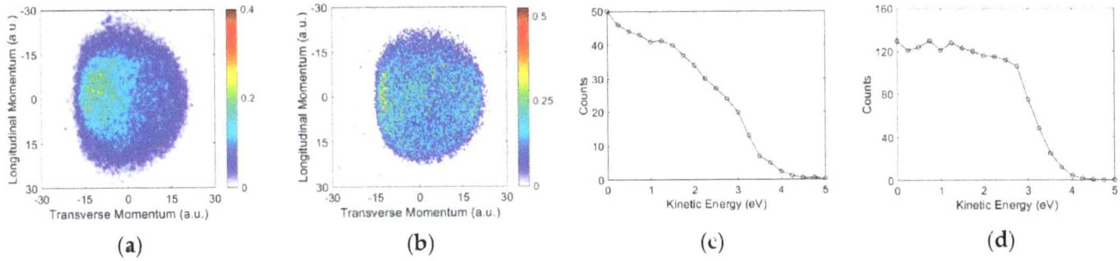

Figure 5. Momentum images obtained for the H^- ions from (**a**) aniline at 8.9 eV and (**b**) benzylamine at 8.4 eV. The incoming electron beam direction is from top to bottom; (**c**,**d**) are the respective kinetic energy distributions.

Figure 6 shows the VSI obtained for the CN^- ion from both molecules. However, although both the molecules show this channel peaking at substantially different electron energies, the momentum images are in the form of a blob. For aniline, the CN^- formation is necessarily associated with the ring-breaking many-body dissociation process, whereas for benzylamine, it can arise from the CH_2NH_2, which is still a many-body break-up mechanism. However, the difference in the resonance energies is indicative of the difference in the origin of this channel. The kinetic energy of this ion in aniline extends up to 0.7 eV peaking at zero. The energy distribution from the benzylamine extends only up to 0.25 eV with a peak at zero. Interestingly, however, the cross-section of this channel in the two molecules differs by more than an order of magnitude.

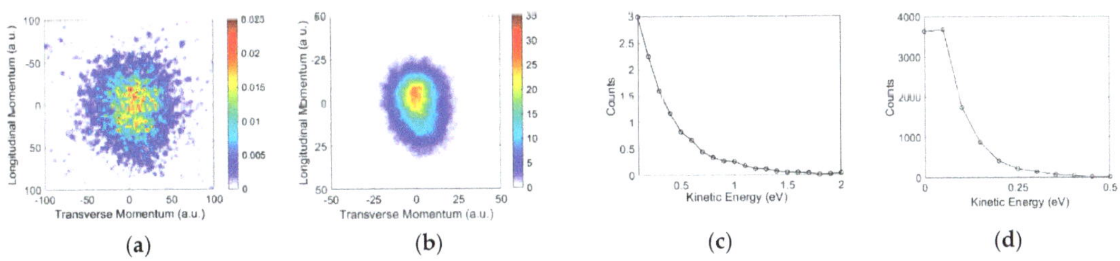

Figure 6. Momentum images obtained for the CN^- ions from (**a**) aniline at 9.6 eV and (**b**) benzylamine at 5.5 eV. The incoming electron beam direction is from top to bottom; (**c**,**d**) are the respective kinetic energy distributions.

4. Conclusions

To conclude, we have shown that the functional group-dependent site-selective fragmentation observed in aliphatic compounds is also observed in aromatic compounds. The original understanding of the site selectivity due to the excitation of the localized electron cloud from the lone pair of electrons holds for the saturated amine where the dissociation dynamics observed in the ammonia are replicated. However, in the unsaturated amine, the delocalized π electron cloud plays a role in the DEA dynamics. The H^- ion signal from the N − H site peaks around 5 eV. This channel results from the two-body dissociation dynamics. The earlier reasoning for the lower absolute cross-section due to the presence of π electron interaction seems to be reasonable as we have observed an order of magnitude difference in the DEA cross-section between aniline and benzylamine. The H^- ion signal from the C − H site peaks between 8 and 9 eV. Like in aliphatic molecules, in aromatic molecules as well, this channel is associated with the many-body fragmentation dynamics.

The CN^- channel, which must result from the many-body dissociation, shows different kinetic energy ranges, possibly due to the difference in its origin in the two channels. It would be interesting to compare the dissociation dynamics of these molecules with benzene and other molecules such as pyridine and pyrrol, where the N atom is part of the aromatic ring.

Author Contributions: Conceptualization, V.S.P. and E.K.; data acquisition, V.T., E.K., S.G. and S.D.; analysis, V.T., S.D. and V.S.P.; interpretation, V.S.P.; writing—original draft preparation, V.S.P.; writing—review and editing, V.S.P. and E.K. All authors have read and agreed to the published version of the manuscript.

Funding: Dept. of Atomic Energy, India, under Project Identification No. RTI4002.

Acknowledgments: E.K. acknowledges the Raja Ramanna Fellowship from the Dept. of Atomic Energy, India.

Conflicts of Interest: The authors declare no conflict of interest.

References

1. Prabhudesai, V.S.; Kelkar, A.H.; Nandi, D.; Krishnakumar, E. Functional Group Dependent Site Specific Fragmentation of Molecules by Low Energy Electrons. *Phys. Rev. Lett.* **2005**, *95*, 143202. [CrossRef]
2. Ptasińska, S.; Denifl, S.; Grill, V.; Märk, T.D.; Illenberger, E.; Scheier, P. Bond- and Site-Selective Loss of H^- from Pyrimidine Bases. *Phys. Rev. Lett.* **2005**, *95*, 093201. [CrossRef] [PubMed]
3. Oster, T.; Kuhn, A.; Illenberger, E. Gas phase negative ion chemistry. *Int. J. Mass Spectrom. Ion Processes* **1989**, *89*, 1–72. [CrossRef]
4. Prabhudesai, V.S.; Nandi, D.; Kelkar, A.H.; Krishnakumar, E. Functional group dependent dissociative electron attachment to simple organic molecules. *J. Chem. Phys.* **2008**, *128*, 154309. [CrossRef] [PubMed]
5. Rescigno, T.N.; Trevisan, C.S.; Orel, A.E.; Slaughter, D.S.; Adaniya, H.; Belkacem, A.; Weyland, M.; Dorn, A.; McCurdy, C.W. Dynamics of dissociative electron attachment to ammonia. *Phys. Rev. A* **2016**, *93*, 052704. [CrossRef]
6. Ram, N.B.; Krishnakumar, E. Dissociative Electron Attachment to Polyatomic Molecules—V: Formic Acid and Propyl Amine. *arXiv* **2010**, arXiv:1007.5169.
7. Skalicky, T.; Allan, M. The assignment of dissociative electron attachment bands in compounds containing hydroxyl and amino groups. *J. Phys. B At. Mol. Opt. Phys.* **2004**, *37*, 4849–4859. [CrossRef]
8. Sanche, L. Nanoscopic aspects of radiobiological damage: Fragmentation induced by secondary low-energy electrons. *Mass Spectrom. Rev.* **2002**, *21*, 349–369. [CrossRef] [PubMed]
9. Gohlke, S.; Illenberger, E. Probing biomolecules: Gas phase experiments and biological relevance. *Europhys. News* **2002**, *33*, 207–209. [CrossRef]
10. Gope, K.; Prabhudesai, V.S.; Mason, N.J.; Krishnakumar, E. Dissociation dynamics of transient anion formed via electron attachment to sulfur dioxide. *J. Chem. Phys.* **2017**, *147*, 054304. [CrossRef]
11. Pikhtovnikov, S.V.; Mavrodiev, V.K.; Furley, I.I.; Gataullin, R.R.; Abdrakhmanov, I.B. Resonance electron capture by aniline molecules and its derivatives. *High Energy Chem.* **2006**, *40*, 224–229. [CrossRef]
12. Rawat, P.; Prabhudesai, V.S.; Rahaman, M.A.; Ram, N.B.; Krishnakumar, E. Absolute cross sections for dissociative electron attachment to NH_3 and CH_4. *Int. J. Mass Spectrom.* **2008**, *277*, 96–102. [CrossRef]
13. King, G.A.; Oliver, T.A.A.; Ashfold, M.N.R. Dynamical insights into $1\pi\sigma*$ state mediated photodissociation of aniline. *J. Chem. Phys.* **2010**, *132*, 214307. [CrossRef] [PubMed]
14. Rajasekhar, B.N.; Veeraiah, A.; Sunanda, K.; Jagatap, B.N. Excited states of aniline by photoabsorption spectroscopy in the 30,000–90,000 cm^{-1} region using synchrotron radiation. *J. Chem. Phys.* **2013**, *139*, 064303. [CrossRef] [PubMed]
15. Ari, T.; Guven, H.; Ecevit, N. Electron energy-loss spectroscopy in monosubstituted benzenes. *J. Electron Spectrosc. Relat. Phenom.* **1995**, *73*, 13–23. [CrossRef]
16. Fenzlaff, H.-P.; Illenberger, E. Low energy electron impact on benzene and the fluorobenzenes: Formation and dissociation of negative ions. *Int. J. Mass Spectrom. Ion Processes* **1984**, *59*, 185–202. [CrossRef]
17. Ram, N.B.; Krishnakumar, E. Dissociative electron attachment resonances in ammonia: A velocity slice imaging based study. *J. Chem. Phys.* **2012**, *136*, 164308. [CrossRef] [PubMed]
18. NIST Chemistry Webbook. Available online: https://webbook.nist.gov/chemistry/ (accessed on 1 September 2022).
19. Bordwell, F.G.; Zhang, X.M.; Cheng, J.P. Bond dissociation energies of the nitrogen-hydrogen bonds in anilines and in the corresponding radical anions. Equilibrium acidities of aniline radical cations. *J. Org. Chem.* **1993**, *58*, 6410–6416. [CrossRef]
20. Luo, Y.R. *Comprehensive Handbook of Chemical Bond Energies*; CRC Press: Boca Raton, FL, USA, 2007.
21. Ram, N.B.; Prabhudesai, V.S.; Krishnakumar, E. Velocity imaging of H^- from formic acid: Probing functional group dependence in dissociative electron attachment. *Eur. Phys. J. D* **2020**, *74*, 49.

atoms

MDPI

Article

Constructing Electron-Atom Elastic Scattering Potentials Using Relativistic Coupled-Cluster Theory: A Few Case Studies

Bijaya Kumar Sahoo

Atomic, Molecular and Optical Physics Division, Physical Research Laboratory, Navrangpura, Ahmedabad 380009, India; bijaya@prl.res.in

Abstract: In light of the immense interest in understanding the impact of an electron on atoms in the low-energy scattering phenomena observed in laboratories and astrophysical processes, we propose an approach to construct potentials using relativistic coupled-cluster (RCC) theory for the determination of electron-atom (e-A) elastic scattering cross-sections (eSCs). The net potential of an electron, scattered elastically by an atom, is conveniently expressed as the sum of the static (V_{st}) and exchange (V_{ex}) potentials due to interactions of the scattered electron with the electrons of the atom and potentials due to polarization effects (V_{pol}) on the scattered electron by the atomic electrons. The V_{st} and V_{ex} potentials for the e-A eSC problems can be constructed with a knowledge of the electron density function of the atom, while the V_{pol} potential can be obtained using the polarizabilities of the atom. In this paper, we present the electron densities and electric polarizabilties of Be, Mg, Ne and Ar atoms using two variants of the RCC method. Using these quantities, we construct potentials for e-A eSC problems. To obtain V_{pol} accurately, we evaluate the second- and third-order electric dipole and quadrupole polarizabilities using a linear response approach.

Keywords: coupled-cluster theory; electron scattering; electric polarizabilities

Citation: Sahoo, B.K. Constructing Electron-Atom Elastic Scattering Potentials Using Relativistic Coupled-Cluster Theory: A Few Case Studies. *Atoms* **2022**, *10*, 88. https://doi.org/10.3390/atoms10030088

Academic Editors: Rajesh Srivastava, Dmitry V. Fursa and Grzegorz Piotr Karwasz

Received: 25 July 2022
Accepted: 1 September 2022
Published: 6 September 2022

1. Introduction

The accurate estimation of scattering cross-sections of electrons with atomic systems is of interest for a wide range of applications in laboratory scattering processes and astrophysics [1–4]. The challenge for the calculation of scattering cross-sections lies in determining accurate wave functions for the scattered electron in the vicinity of an atomic target [5,6]. The coupling between the scattered wave functions and atomic wave functions are addressed through the close-coupling [7] and R-matrix [8] formalism, but they are mostly used in a non-relativistic framework [9,10] owing to the complexity involved in the relativistic formalism. In another approach, the interactions among the scattered electron and atomic electrons are included by splitting them into two parts—an electron-electron correlation component and the effects of electron polarization due to the atomic electrons [11–16]. In this approach, the wave functions of the electron and atom are solved separately. The electron correlation effects within the atom are accommodated via a suitable many-body method in the determination of the atomic wave functions (equivalent to atomic wave density functions (ρ)). These functions are further used to construct the interaction potential for the scattered electron. It has both direct and exchange terms owing to the indistinguishable nature of the electrons. An atom is polarized due to the charged scattered electron which modifies the behavior of its wave functions. This effect also influences the construction of the effective potentials of the scattered electrons and is estimated using the electric polarizabilities of the atom. These effective potentials are used to obtain the wave functions of the scattered electrons, for different ranges of kinetic energies, using a distorted wave function (DW) formalism [17,18]. For a highly energetic scattered electron, it is desirable to use the relativistic Dirac equation in the DW approximation (RDW method) [19–22].

In light of several applications of electron-atom scattering cross-sections, such as in modelling metal vapour lasers and plasma plasma environments [23], insights have been gained into different physical processes in many natural and technological environments, including the Earth's atmosphere and in the atmospheres of other planets and their satellites [24], as well as understanding of electron-atom interactions [25]. Theoretical studies on electron scattering by Be, Mg, Ca, Ne, Ar and other atoms have previously been carried out [26–33]. Most of these atoms have closed-shell electronic configurations.

From the above discussion, it is clear that improvement in the accuracy of the scattering cross-section depends on the accurate evaluation of the atomic wave function and the electric polarizabilities of the atom. Typical many-body methods employed to determine the atomic wave functions include a method informed by many-body perturbation theory (MBPT method), the configuration interaction (CI) method, and the coupled-cluster (CC) method. Among these, the CC method is viewed as the "gold standard" for its ability to incorporate electron correlations in the determination of the atomic wave functions at a given approximation level [34–37]. Here, we employ the CC method in a relativistic framework (RCC method) to evaluate the atomic wave functions. Although the (R)CC method has previously been widely applied to calculate many spectroscopic properties to a high degree of accuracy, its ability to obtain scattering cross-sections has not been rigorously tested, except in our first demonstrations in Mg^+ [38] and Ca [39] when studying scattering cross-sections in plasma embedded and confined atom problems. Furthermore, atomic polarization effects on the scattering cross-sections are quite significant. Often, only contributions from the electric dipole polarizabilities (α_d) are considered in the construction of scattering potentials due to their dominant contributions. In recent calculations, it has been shown that contributions arising through the electric quadrupole (α_q) and coupled dipole-quadrupole (B) polarizabilities are non-negligible [15,40]. The aim of the present investigation was to provide general approaches to accurately determine the ρ, α_d, α_q and B values of atomic systems by employing an RCC method that can be used whenever required to obtain the elastic scattering cross-sections of an electron from the closed-shell atomic systems. For representation purposes, we give the results for Be, Ne, Mg and Ar atoms; however, the scheme is very general and can be extended to atomic systems with open-shell configurations.

Apart from the application of electric polarizabilities to determine electron scattering potentials, they are also immensely important for estimating Stark shifts in atomic energy levels. This is why atomic polarizability studies are interesting in their own right. In the literature, α_d has been extensively studied due to its predominant contribution to the energy shift, followed by α_q then B in the presence of an external electric field. Recently, we proposed a linear response approach to determine the α_d, α_q and B values for Zn in the RCC and relativistic normal CC (RNCC) theory frameworks [41]. We had found that the results from the RCC and RNCC theories differed significantly in the commonly considered singles and doubles approximation. Here, we investigate ρ, α_d, α_q and B values using both methods, and compare them with previously reported results for Be, Ne, Mg and Ar atoms. Using these values, we determine the electron scattering potentials and represent these by plotting them against the radial distances. Though these potentials are obtained using a relativistic method, the estimated potentials can be used in both the DW and RDW methods to calculate electron scattering cross-sections with different projectile energies.

2. Theory

For the spherically symmetric interaction potential $V(r)$ of a projectile electron with the target, the direct and exchange scattering amplitudes can be determined by [42]

$$f(k,\theta) \quad = \quad \frac{1}{2\iota k}\sum_{l=0}^{\infty}((l+1)(\exp(2\iota\delta_{\kappa=-l-1})-1)+l(\exp(2\iota\delta_{\kappa=l})-1))P_l(\cos\theta) \quad (1)$$

and

$$g(k,\theta) = \frac{1}{2\imath k}\sum_{l=0}^{\infty}(\exp(2\imath\delta_{\kappa=l}) - \exp(2\imath\delta_{\kappa=-l-1}))P_l^1(\cos\theta) \quad (2)$$

Here k is the relativistic wave number, $\delta_{\kappa=-l-1,l}$ are the scattering phase shifts with $\kappa = -l-1$ and $\kappa = l$ refer to the relativistic quantum numbers for projectile electron with $j = l + 1/2$ and $j = l - 1/2$, respectively. In the above equation, θ is the scattering angle, and $P_l(\cos\theta)$ and $P_l^1(\cos\theta)$ are Legendre polynomials and associated Legendre functions, respectively. Using these amplitudes, the differential cross-sections per unit solid angle for spin unpolarized electrons can be calculated by

$$\frac{d\sigma}{d\Omega} = |f(k,0)|^2 + |g(k,0)|^2, \quad (3)$$

from which integrated cross-sections can be estimated by integrating over the solid angle. In the (R)DW approximation, the first-order scattering amplitude of an electron from an atomic system with nuclear charge Z and N number of electrons can be expressed as

$$f(J_f,\mu_f;J_i,\mu_i,\theta) = 4\pi^2\sqrt{\frac{k_f}{k_i}}\langle F_{DW}^{k_f}|H_{scat}|F_{DW}^{k_i}\rangle, \quad (4)$$

where J and μ represent the angular momenta of the states of the atomic target and the scattered electron, respectively, k is the momentum of the scattered electron and F_{DF} are the (R)DW wave functions, while the subscript i denotes the initial state and f denotes the final state. A similar expression can be given for g. In the DW method, the effective scattering Hamiltonian in atomic units (a.u.) is given by

$$H_{scat} = -\frac{1}{2}\nabla^2 + V(r) \quad (5)$$

whereas in the RDW method, it is given by

$$H_{scat} = c\alpha \cdot \mathbf{p} + \beta c^2 + V(r). \quad (6)$$

Here c is the speed of light, α and β are the Dirac matrices and $V(r)$ is the scattering potential. For accurate determination of scattering cross-sections, it is imperative to obtain $V(r)$ accurately. In a more convenient form, $V(r)$ can be expressed as [11]

$$V(r) = V_{st}(r) + V_{ex}(r) + V_{pol}(r), \quad (7)$$

where $V_{st}(r)$, $V_{ex}(r)$ and $V_{pol}(r)$ are known as the static, exchange and polarization potentials, respectively. The static potential can have contributions from the nuclear potential ($V_{nuc}(r)$) and the direct electron-electron Coulomb interaction potential $V_C(r)$; i.e., $V_{st}(r) = V_{nuc}(r) + V_C(r)$. Usually, a point-like atomic nucleus is considered in the scattering cross-section calculations by defining $V_{nuc}(r) = -\frac{Z}{r}$ for the atomic number of the system Z. In the present study, we have used the Fermi-charge distribution, given by

$$\rho_A(r) = \frac{\rho_0}{1 + e^{(r-c)/a}}, \quad (8)$$

where ρ_0 is the normalization constant, c is the half-charge radius and $a = 2.3/4ln(3)$ is known as the skin thickness, to take into account the finite size effect of the nucleus. This corresponds to the expression for the nuclear potential, as [43]

$$V_{nuc}(r) = -\frac{Z}{\mathcal{N}r} \times$$
$$\begin{cases} \frac{1}{c}(\frac{3}{2} + \frac{a^2\pi^2}{2c^2} - \frac{r^2}{2c^2} + \frac{3a^2}{c^2}P_2^+ \frac{6a^3}{c^2 r}(S_3 - P_3^+)) & \text{for } r_i \leq c \\ \frac{1}{r_i}(1 + \frac{a62\pi^2}{c^2} - \frac{3a^2 r}{c^3}P_2^- + \frac{6a^3}{c^3}(S_3 - P_3^-)) & \text{for } r_i > c, \end{cases} \tag{9}$$

where the factors are

$$\begin{aligned} \mathcal{N} &= 1 + \frac{a^2\pi^2}{c^2} + \frac{6a^3}{c^3}S_3 \\ \text{with } S_k &= \sum_{l=1}^{\infty} \frac{(-1)^{l-1}}{l^k} e^{-lc/a} \\ \text{and } P_k^{\pm} &= \sum_{l=1}^{\infty} \frac{(-1)^{l-1}}{l^k} e^{\pm l(r-c)/a}. \end{aligned} \tag{10}$$

Similarly, we can express $V_C(r) = \sum_{b=1}^{N_e} \langle \phi_b | \frac{1}{|\vec{r} - \vec{r}_b|} | \phi_b \rangle$ with N_e, denoting the total number of electrons of the target atom, and $|\phi_b\rangle$ is the single particle wave function of the atomic electron b such that

$$\frac{1}{|\vec{r}_i - \vec{r}_j|} = \sum_{k=0}^{\infty} \frac{4\pi}{2k+1} \frac{r_<^k}{r_>^{k+1}} \sum_{q=-k}^{k} Y_q^{k*}(\theta, \varphi) Y_q^k(\theta, \varphi), \tag{11}$$

where $r_> = \max(r_i, r_j)$, $r_< = \min(r_i, r_j)$, and $Y_q^k(\theta, \varphi)$ is the spherical harmonics of rank k with its component q. In terms of the Racah operator (C_q^k), the above expression is given by a scalar product as

$$\frac{1}{|\vec{r}_i - \vec{r}_j|} = \sum_{k=0}^{\infty} \frac{r_<^k}{r_>^{k+1}} \mathbf{C}^k(\hat{r}_i) \cdot \mathbf{C}^k(\hat{r}_j). \tag{12}$$

In the Dirac theory, the single particle orbital wave functions are given by

$$|\phi(r)\rangle = \frac{1}{r} \begin{pmatrix} P(r)\chi_{jm_j l_L}(\theta, \varphi) \\ iQ(r)\chi_{jm_j l_S}(\theta, \varphi) \end{pmatrix}, \tag{13}$$

where the upper and lower components are the large and small components of the single particle wave function, respectively, $P(r)$ and $Q(r)$ denote the radial parts of these components, and the χ's denote the spin angular parts of each component which depend on the quantum numbers j, m_j, and l. l_L denotes l for the large component, while l_S denotes l for the small component. Thus, for a closed-shell atomic target, such as those under consideration here, we can have

$$V_C(r) = \sum_b (2j_b + 1) \int_0^{\infty} dr_b \frac{1}{r_>} \left[P_b^2(r_b) + Q_b^2(r_b)\right]. \tag{14}$$

It is worth noting that, for open-shell atomic targets, there will be a finite value of multipoles k in the above expression and the computation of $V_C(r)$ will be quite difficult, but is possible [38]. Using density function formalism, the above expression can be given by

$$V_C(r) = \sum_b \left[\frac{1}{r} \int_0^r dr_b \rho_b(r_b) r_b'^2 + \int_r^{\infty} dr_b \rho_b(r_b) r_b\right], \tag{15}$$

where the atomic density function is given by $\hat{\rho}(r) = \sum_i \hat{\rho}_i(r) = \sum_i |\phi_i\rangle\langle\phi_i|$ with

$$\langle \phi_j | \hat{\rho}_i(r) | \phi_k \rangle = \delta_{ji}\delta_{ik}(P_j(r)P_k(r) + Q_j(r)Q_k(r)). \tag{16}$$

It is not possible to determine $V_{ex}(r)$ separately as it depends on the wave function of the scattered electron itself. However, it can be approximately estimated using the Hara free electron gas model, given by [44]

$$V_{ex}(r) = -\frac{2}{\pi}K_F(r)F[\eta(r)], \tag{17}$$

where the Fermi momentum $K_F(r) = (3\pi\rho(r))^{1/3}$ and $F(\eta) = \frac{1}{2} + \frac{1-\eta^2}{4\eta}ln\left|\frac{1+\eta}{1-\eta}\right|$ with $\eta(r) = \frac{K(r)}{K_F(r)}$ for the local electron momentum given by

$$K^2(r) = K_F^2 + 2I + k^2. \tag{18}$$

Here, I denotes the ionization potential (IP) of the target atom and $k^2/2$ is the kinetic energy of the projectile electron. This means that evaluation of $V_{ex}(r)$ requires the atomic density function and the IP of the atom, along with the kinetic energy of the projectile. Since the kinetic energy of the projectile is arbitrary, we provide here only the $\rho(r)$ values, while IPs can be used from the experimental data.

The polarization potential is given by [45,46]

$$V_{pol}(r) = -\left(\frac{\alpha_d}{2r^4} + \frac{\alpha_q}{2r^6} - \frac{B}{2r^7} + O(1/r^8)\right) \times \left[1 - e^{(r/r_c)^6}\right], \tag{19}$$

where α_d, α_q and B are known as the second-order dipole, second-order quadrupole and third-order dipole-quadrupole polarizabilities, respectively. $O(1/r^8)$ corresponds to the higher-order polarizability contributions and is neglected here. r_c is an adjustable parameter, which can be determined by estimating IP using the above potential in the equation of motion, and is assumed to be different for different atoms and also for different levels of approximation in the above expression. For convenience and demonstration purposes, without losing much accuracy, we have considered $r_c = 3.5$ in atomic units (a.u.) for all the atoms considered [46].

In the following section, we present the RCC method to estimate $\rho(r)$, V_{st}, α_d, α_q and B in the closed-shell atomic systems. In place of calculating $V_{st}(r)$ directly using RCC theory, we estimate it by evaluating $V_{nuc}(r)$ and $V_C(r)$ separately with $V_C(r)$ obtained from the $\rho(r)$ values. The expectation values of the operators are again evaluated using the standard RCC and RNCC theory frameworks, and the results are compared with the earlier reported literature values.

3. Methods for Calculations

Since α_d, α_q and B are determined by treating electric dipole operator D and quadrupole operator Q as external perturbations, the atomic wave functions without these external operators are denoted with the superscript 0 ($|\Psi_0^{(0)}\rangle$). We have utilized the Dirac–Coulomb Hamiltonian to determine these unperturbed wave functions, given by

$$H_0 = \sum_{i=1}^{N_e}\left[c\alpha_i \cdot \mathbf{p}_i + (\beta_i - 1)c^2 + V_{nuc}(r_i) + \sum_{j>i}\frac{1}{r_{ij}}\right], \tag{20}$$

where $r_{ij} = |\vec{r}_i - \vec{r}_j|$ is the inter-electronic separation between the electrons located at the r_i and r_j radial positions with respect to the center of the nucleus.

The density matrix of the atomic state $|\Psi_0^{(0)}\rangle$ can be determined by

$$\rho(r) = \frac{\langle\Psi_0^{(0)}|\hat{\rho}(r)|\Psi_0^{(0)}\rangle}{\langle\Psi_0^{(0)}|\Psi_0^{(0)}\rangle}. \tag{21}$$

Following [41], the expressions for α_d, α_q and B of the ground state of a closed-shell system can be given by

$$\alpha_d = 2\frac{\langle \Psi_0^{(0)}|D|\Psi_0^{(d,1)}\rangle}{\langle \Psi_0^{(0)}|\Psi_0^{(0)}\rangle},$$

$$\alpha_q = 2\frac{\langle \Psi_0^{(0)}|Q|\Psi_0^{(q,1)}\rangle}{\langle \Psi_0^{(0)}|\Psi_0^{(0)}\rangle} \quad (22)$$

and

$$B = 2\frac{\langle \Psi_0^{(d,1)}|D|\Psi_0^{(q,1)}\rangle}{\langle \Psi_0^{(0)}|\Psi_0^{(0)}\rangle}, \quad (23)$$

where $|\Psi_0^{(0)}\rangle$ and $|\Psi_0^{(o,1)}\rangle$ are the zeroth-order wave function and the first-order wave function of the atom due to an operator $O \equiv D$ or Q.

From the above expressions, it is clear that accurate evaluations of α_d, α_q and B depend on the many-body method employed to determine $|\Psi_0^{(0)}\rangle$ and $|\Psi_0^{(o,1)}\rangle$. These wave functions can be determined by solving the following equations

$$H_0|\Psi_0^{(0)}\rangle = E_0^{(0)}|\Psi_0^{(0)}\rangle \quad (24)$$

and

$$(H_0 - E_0^{(0)})|\Psi_0^{(o,1)}\rangle = (E_0^{(o,1)} - O)|\Psi_0^{(0)}\rangle \quad (25)$$

with the first-order energy correction $E_0^{(o,1)}$ due to O, which is zero in the present study.

Our intention here is to demonstrate the evaluation of $\rho(r)$, α_d, α_q and B in the closed-shell atoms using the RCC and RNCC theories to construct the electron-atom scattering potentials. In the RCC theory, we can express [47,48]

$$|\Psi_0^{(0)}\rangle = e^{T^{(0)}}|\Phi_0\rangle, \quad (26)$$

and

$$|\Psi_0^{(o,1)}\rangle = e^{T^{(0)}}T^{(o,1)}|\Phi_0\rangle, \quad (27)$$

where $T^{(0)}$ accounts for electron correlation effects, and $T^{(o,1)}$ includes electron correlations, along with the effect due to O, while acting on the Dirac–Hartree–Fock (DHF) wave function $|\Phi_0\rangle$ of the system.

In this approach, the expressions for ρ, α_d, α_q and B are given by [41]

$$\rho(r) = \frac{\langle\Phi_0|e^{T^{(0)\dagger}}\hat{\rho}(r)e^{T^{(0)}}|\Phi_0\rangle}{\langle\Phi_0|e^{T^{(0)\dagger}}e^{T^{(0)}}|\Phi_0\rangle}, \quad (28)$$

$$\alpha_d = 2\frac{\langle\Phi_0|e^{T^{(0)\dagger}}De^{T^{(0)}}T^{(d,1)}|\Phi_0\rangle}{\langle\Phi_0|e^{T^{(0)\dagger}}e^{T^{(0)}}|\Phi_0\rangle}, \quad (29)$$

$$\alpha_q = 2\frac{\langle\Phi_0|e^{T^{(0)\dagger}}Qe^{T^{(0)}}T^{(q,1)}|\Phi_0\rangle}{\langle\Phi_0|e^{T^{(0)\dagger}}e^{T^{(0)}}|\Phi_0\rangle} \quad (30)$$

and

$$B = 2\frac{\langle\Phi_0|T^{(d,1)\dagger}e^{T^{(0)\dagger}}De^{T^{(0)}}T^{(q,1)}|\Phi_0\rangle}{\langle\Phi_0|e^{T^{(0)\dagger}}e^{T^{(0)}}|\Phi_0\rangle}. \quad (31)$$

Evaluating the above expressions involves two major challenges, even after making approximations in the level of excitations in the RCC calculations. The first being that there are two non-terminating series in the numerator and denominator. The second challenge is that the numerator can have factors which are both connected and disconnected to the operators D or Q. These problems can be partially addressed by defining a normal-order form of operators with respect to $|\Phi_0\rangle$, in which the above expressions can be simplified to [49,50]

$$\rho(r) = \langle\Phi_0|e^{T^{(0)\dagger}}\hat{\rho}(r)e^{T^{(0)}}T^{(d,1)}|\Phi_0\rangle_c, \tag{32}$$

$$\alpha_d = 2\langle\Phi_0|e^{T^{(0)\dagger}}De^{T^{(0)}}T^{(d,1)}|\Phi_0\rangle_c, \tag{33}$$

$$\alpha_q = 2\langle\Phi_0|e^{T^{(0)\dagger}}Qe^{T^{(0)}}T^{(q,1)}|\Phi_0\rangle_c \tag{34}$$

and

$$B = 2\langle\Phi_0|T^{(d,1)\dagger}e^{T^{(0)\dagger}}De^{T^{(0)}}T^{(q,1)}|\Phi_0\rangle_c, \tag{35}$$

where subscript c denotes connected terms only appearing within the respective expression. Although this removes the non-terminating series appearing in the denominator, it still contains a non-terminating series in the numerator. Further, the above expressions with connected terms hold good only when there is no approximation made in the T operator. In practice, T is truncated as for our RCCSD method. These expressions again do not satisfy the Hellman–Feynman theorem [34]. All these problems can be circumvented by the RNCC theory.

In the RNCC theory, the ket state is the same as in RCC theory but the bra state is replaced by

$$\langle\tilde{\Psi}^{(0)}| = \langle\Phi_0|(1+\Lambda^{(0)})e^{-T^{(0)}}, \tag{36}$$

with a de-excitation operator $\Lambda^{(0)}$ that satisfies

$$\langle\tilde{\Psi}^{(0)}|\Psi^{(0)}\rangle = \langle\Phi_0|(1+\Lambda^{(0)})e^{-T^{(0)}}e^{T^{(0)}}|\Phi_0\rangle = 1. \tag{37}$$

It can be shown that the eigenvalues of both $\langle\Psi^{(0)}|$ and $\langle\tilde{\Psi}^{(0)}|$ are the same if

$$\langle\Phi_0|\Lambda\bar{H}_0|\Phi_0\rangle = 0, \tag{38}$$

where $\bar{H} = e^{-T^{(0)}}H_0e^{T^{(0)}} = (He^T)_c$.

Now, we can write the first-order perturbed wave function in the RNCC theory as [41,51]

$$\langle\tilde{\Psi}^{(o,1)}| = \langle\Phi_0|\left[\Lambda^{(o,1)} + (1+\Lambda^{(0)})T^{(o,1)}\right]e^{-T^{(0)}}. \tag{39}$$

Consequently, the RNCC expressions for $\rho(r)$, α_d, α_q and B are given by

$$\rho(r) = \langle\Phi_0|\left(1+\Lambda^{(0)}\right)\tilde{\hat{\rho}}(r)|\Phi_0\rangle, \tag{40}$$

$$\alpha_d = \langle\Phi_0|\left(1+\Lambda^{(0)}\right)\tilde{D}T^{(d,1)} + \Lambda^{(d,1)}\tilde{D}|\Phi_0\rangle, \tag{41}$$

$$\alpha_q = \langle\Phi_0|\left(1+\Lambda^{(0)}\right)\tilde{Q}T^{(q,1)} + \Lambda^{(q,1)}\tilde{Q}|\Phi_0\rangle \tag{42}$$

and

$$B = \langle\Phi_0|\Lambda^{(d,1)}\tilde{D}T^{(q,1)} + \Lambda^{(q,1)}\tilde{D}T^{(d,1)}|\Phi_0\rangle, \tag{43}$$

where $\tilde{O} = (Oe^{T^{(0)}})_c$. In the RNCC theory, we also consider only the singles and doubles excitations (RNCCSD method) to carry out the calculations. It is worth noting here that the next leading-order electron correlation effects to $\rho(r)$, α_d, α_q and B arising through the higher-level excitations will converge faster in the RNCC theory than the RCC theory [41,51].

4. Results and Discussion

We first evaluate the density functions $\rho(r)$ of the ground states of the Be, Mg, Ne and Ar atoms. Since the correlation contributions, i.e.,the differences between the DHF and RCC/RNCC values (given as $\delta\rho(r)$), to these functions are very small compared to the DHF values, we consider these contributions separately. In Figure 1, we plot $\rho(r)$ values from the DHF method, while the correlation contributions $\delta\rho(r)$ from the RCCSD and RNCCSD methods are shown in Figure 2. As can be seen from the first figure, the density profiles of Be, Mg, Ne and Ar appear to be different. This suggests that the electronic charge distributions among these atoms are quite different. From the second figure, we see that there are slight differences in the correlation contributions from the RCCSD and RNCCSD methods in Be, while, for the other atoms, not much difference is observed. As mentioned earlier, accurate values of α_d, α_q and B are important in determining $V_{pol}(r)$ for the electron-atom scattering problem. Therefore, the roles of the electron correlation effects through the RCCSD and RNCCSD methods in the above atoms can be better understood through the calculation of electric polarizabilities.

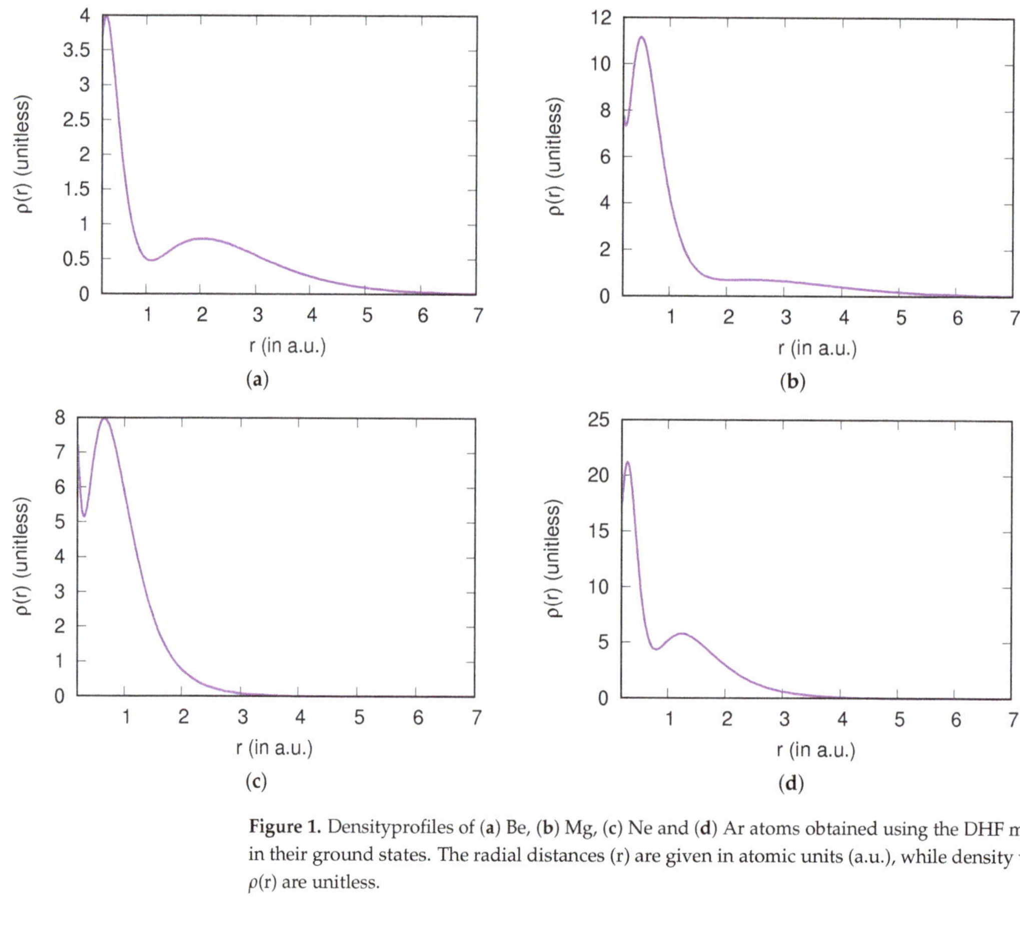

Figure 1. Densityprofiles of (**a**) Be, (**b**) Mg, (**c**) Ne and (**d**) Ar atoms obtained using the DHF method in their ground states. The radial distances (r) are given in atomic units (a.u.), while density values ρ(r) are unitless.

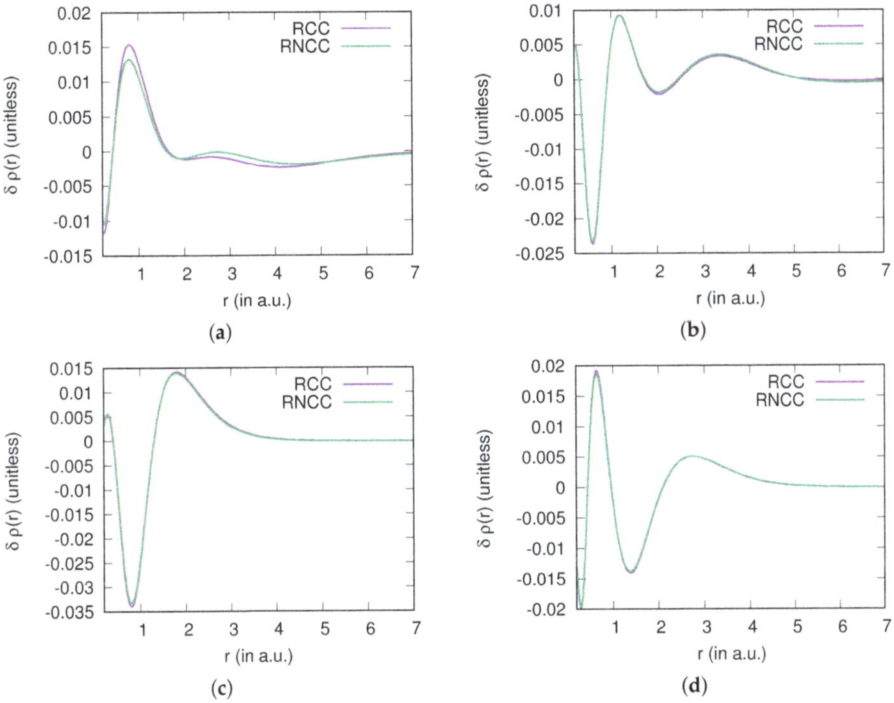

Figure 2. Correlation contributions to $\rho(r)$ (shown in figure as $\delta\rho(r)$) from the RCC and RNCC methods in (**a**) Be, (**b**) Mg, (**c**) Ne and (**d**) Ar. As seen, the $\delta\rho(r)$ values are almost the same through the RCC and RNCC methods in all atoms except Be in which slight differences are evident.

To our knowledge, there are no available calculations of $\rho(r)$ of the atoms considered which explicitly use the (R)CC methods previously discussed. In a recent study [52], a CI method was employed in a non-relativistic framework to determine density functions for studying the quantum potential neural network of Li, Be and Ne atoms. We found that the density function behaviors we obtained for Be and Ne almost matched those of the density functions of these atoms reported in [52]. We could not find any reference which specifically reported the density functions of Mg and Ar; however, from analyses of radial function distributions in Ne and Ar shown in [53], we assume that the behavior of the density functions of the Ar atom we obtained using the DHF method follow the correct trend. Moreover, in a different investigation [54], calculations of the $\rho(r)$ values in carbon atoms followed similar trends to our results for Mg. From all these analyses, we infer that our $\rho(r)$ values for Mg should be correct. Since the previous studies did not explicitly discuss $\delta\rho(r)$ contributions, we were unable to compare our findings for these values with any other calculations.

In Table 1, we present the α_d, α_q and B values calculated using the DHF, RCCSD and RNCCSD methods. It can be seen from this table that there are large differences between the results from the DHF and RCCSD methods. These differences become larger in the determination of α_q followed by the B values. The RCCSD values of B in the alkaline-earth atoms are about 2.5 times larger than the DHF values. In all atoms, the RNCCSD values of α_d, α_q and B are seen to be lower than the RCCSD values, except in the determination of α_q in the Be atom. The α_d values from the RCCSD and RNCCSD methods are very close to each other, but there are significant differences observed among the α_q values of the RCCSD and RNCCSD methods. These differences are quite prominent in the evaluation of the B values. As discussed in the previous section, an approximated RNCC method is more reliable in the determination of properties than an approximated RCC method; thus, we believe that our RNCCSD results are more accurate and should be treated here as the final results.

Table 1. Our calculated values of α_d, α_q and B (in a.u.) of the Be, Mg, Ne and Ar atoms from the DHF, RCCSD and RNCSSD methods. These values are also compared with precise values from the literature.

Property	DHF	RCCSD	RNCCSD	Others
		Be atom		
α_d	30.53	38.33	37.40	37.739(30) [55] 37.76(22) [56] 37.86(17) [57] 37.74(3) [58]
α_q	220.15	299.82	304.34	300.96 [55] 300.6(3) [56]
B	−1218.38	−2729.17	−2172.95	−2100(60) [55]
		Mg atom		
α_d	54.94	71.74	69.40	71.22(36) [55] 71.3(7) [56] 72.54(50) [57] 71.2(4) [58]
α_q	567.37	809.56	797.91	813.9(16.3) [55] 812(6) [56]
B	−3847.89	−9293.74	−7226.24	−7750(780) [55]
		Ne atom		
α_d	1.98	2.70	2.62	2.6669(8) [59] 2.652(15) [57] 2.66110(3) [58] 2.64 [60]
α_q	4.76	7.48	7.09	7.52(15) [55] 7.36 [60]
B	−6.15	−14.38	−11.67	−18.12(54) [55] −17.27 [60]
		Ar atom		
α_d	10.15	11.21	11.15	11.083(7) [61] 11.070(7) [62] 11.089(4) [57] 11.083(7) [58] 11.33 [63] 10.73 [64]
α_q	37.19	51.61	50.33	53.37(1.07) [55] 53.22 [63] 49.46 [64]
B	−71.07	−140.53	−115.35	−159(8) [55] −167.5 [63] −141 [64]

Due to the enormously wide use of electric polarizabilities in various experimental applications, a number of theoretical calculations have been presented in the literature. We consider the results obtained from previous experiments [59,61,62], sources that provide compilations of earlier data [55,58], our own previous RCC calculations [57], and evidence cited in papers that report most of these quantities using a single many-body method [56,60,63,64]. The results of other studies are mostly summarized in [55,58]. Many earlier theoretical studies determined the α_d values, with less theoretically based results found for the α_q values of the atoms considered. To our knowledge, only a few non-relativistic calculations for the B values of the Be, Mg, Ne and Ar atoms considered have been reported [55,60,63,64]. Furthermore, we did not find any experimental results of α_d for Be and Mg, but precisely measured α_d values are available for Ne and Ar. Both of our RCCSD and RNCSSD values are in agreement with those obtained from previous calculations. We note that our RCCSD value of α_q is closer to the previously reported precise calculation result than the RNCCSD value; however, this trend is different for the

α_d and B values. These findings are slightly different for the Mg atom, where it is observed that both the α_d and α_q values from our RCCSD method closely matched the previously reported accurate calculations, but the RNCCSD value for B was in closer agreement with the previous calculation [55]. From these comparisons, it is not possible to argue that the RCCSD method values are more accurate than the RNCCSD results wherever they agree with the previous calculations unless they are verified experimentally. The α_q and B values, previously estimated using a finite-field (FF) approach, are not numerically reliably. The experimental value of α_d in Ne is very precise, and comparison of theoretical results with this value can indicate the validity of the many-body methods employed. We also compared our RNCCSD values of α_d and α_q with the literature values in Table 1. For Be, there are several calculations of α_d available; we have listed some of the precise theoretical results in the above table from the CC and RCC calculations. Several calculations of α_q of the considered atoms, including Be, have been reported using non-relativistic variation-perturbation methods using a finite-field (FF) approach [55,60,63,64], and using a combined CI and MBPT (CI + MBPT) method in a sum-over-states approach [56]. Our RNCCSD α_d value closely matches the previously estimated values. We found a slight difference for the α_q value from the RNCCSD method and the previously reported precise value using the CI + MBPT method [56]. Our RCCSD value of α_d in Mg agrees closely with the previously calculated values using various many-body methods. However, the previously reported α_d values from different calculations spread over a wide-range. This is due to the large electron correlation effects exhibited by both the valence electrons of the Mg atom. Nonetheless, our RNCCSD value of α_d is also close to that obtained from other calculations. However, our RCCSD value for α_q is closer to previous calculations while the RNCCSD result differs significantly from the earlier calculations. From this difference, we cannot say with confidence that the RCCSD value is more accurate than the RNCCSD result. This is because the earlier predicted α_q values are obtained using non-relativistic methods or lower-order relativistic methods. Thus, only empirical measurements can confirm the reliability of these calculations. Comparing the α_d value of Ne with experiment [59], our RCCSD value is closer to the experimental value than the RNCCSD value. We anticipate that after including Breit and quantum electrodynamic corrections, the RNCCSD value will improve further. Similarly, the α_q value from the RCCSD method is closer to the previous calculations than the RNCCSD method. Since there is no experimental result for α_q available, we cannot claim that the RNCCSD value is less accurate than the RCCSD result. Similar trends for the α_d and α_q values can be seen in the Ar atom.

Compared to the α_d and α_q values, B values have received little attention both in theoretical and experimental studies. The contributions of these values to the Stark effects are extremely small when being precisely observed. Strong electron correlation effects are also involved when evaluating B values accurately. In addition, extrapolation of B values from the FF approach requires inclusion of both the electric dipole and quadrupole field interactions in the atomic Hamiltonian. In the linear response approach, estimations of the B values demand calculation of first-order perturbed wave functions due to both the electric dipole and quadrupole operators. These are the main reasons why the B values are not widely investigated in many atomic systems. We identified some literature values for B of the Be, Mg, Ne and Ar atoms [55,60,63,64] which are listed in Table 1. These literature values are obtained by adopting an FF approach in a non-relativistic framework. By comparison of our calculations with literature values, we note that our RNCCSD values agree with the earlier reported values, while the RCCSD results differ greatly for both the Be and Mg atoms. However, this is reversed for the Ne and Ar atoms. The reason for this could be that different many-body methods were considered to estimate the B values of the alkaline-earth atoms and of the noble gas atoms. We assume that our RNCSSD results are more reliable compared to all the listed values in Table 1.

In Figure 3a,b, we show the individual contributions to $V(r)$ from the RCCSD and RNCCSD methods for the Be atom. As can be seen in both these plots, the contributions from V_{nuc} dominates while the V_C contributions are also quite visible. There are also noticeable contributions arising from the $V_{pol}(r)$. Similar trends can also be observed for the RCCSD values, as shown in Figure 3c, and the RNCCSD values, as shown in Figure 3d, for the Mg atom, but the shapes are slightly different due to $V_C(r)$ and $V_{pol}(r)$ contributions. These differences can be understood from the density profiles of both the atoms shown in Figure 1. In Figure 4a–d, we show different contributions to $V(r)$ from the RCCSD and RNCCSD methods for Ne and Ar. As can be seen from the figure, the trends from individual contributions to $V(r)$ in both Ne and Ar appear quite similar except for their magnitudes. The figure also shows that contributions from $V_{pol}(r)$ are negligibly small in both the atoms. Compared to the alkaline-earth atoms, the results for both the Ne and Ar atoms look quite similar to those for the Mg atom. It is of note that the density profiles shown in Figure 1 of the Be and Ar atoms appear similar, while the density profiles of Mg and Ne appear to have similar features. Thus, it is not possible to obtain a clear picture of the scattering potential behavior of an electron from an atom just by looking at the density profile of the atom. Nonetheless, we have discussed procedures to construct the electron-atom scattering potentials by evaluating contributions from the static and polarization potentials due to the Be, Mg, Ne and Ar atoms using the RCC and RNCC methods. These procedures can also be adopted for heavier closed-shell atomic systems, where electron correlation effects could be very pronounced.

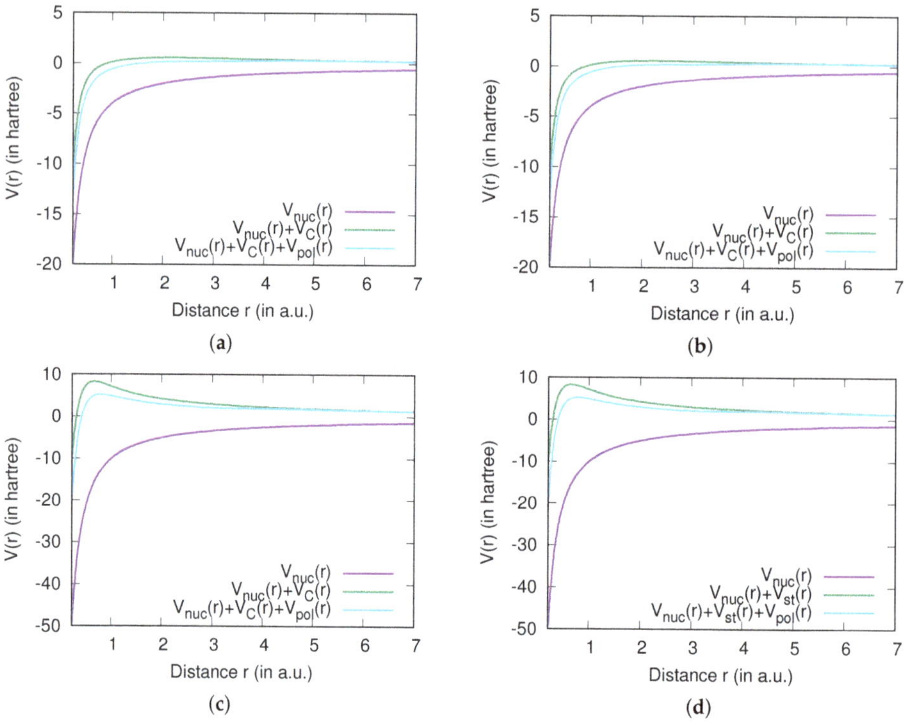

Figure 3. Plots demonstrating comparative analyses of contributions from $V_{nuc}(r)$, $V_C(r)$ and $V_{pol}(r)$ to the electron scattering potential $V(r)$ from the Be and Mg alkaline-earth atoms. In (**a**,**b**), results are given from the RCCSD and RNCCSD methods, respectively, for the Be atom. Results from the RCCSD and RNCCSD methods are shown in (**c**,**d**), respectively, for the Mg atom. All quantities are given in a.u.

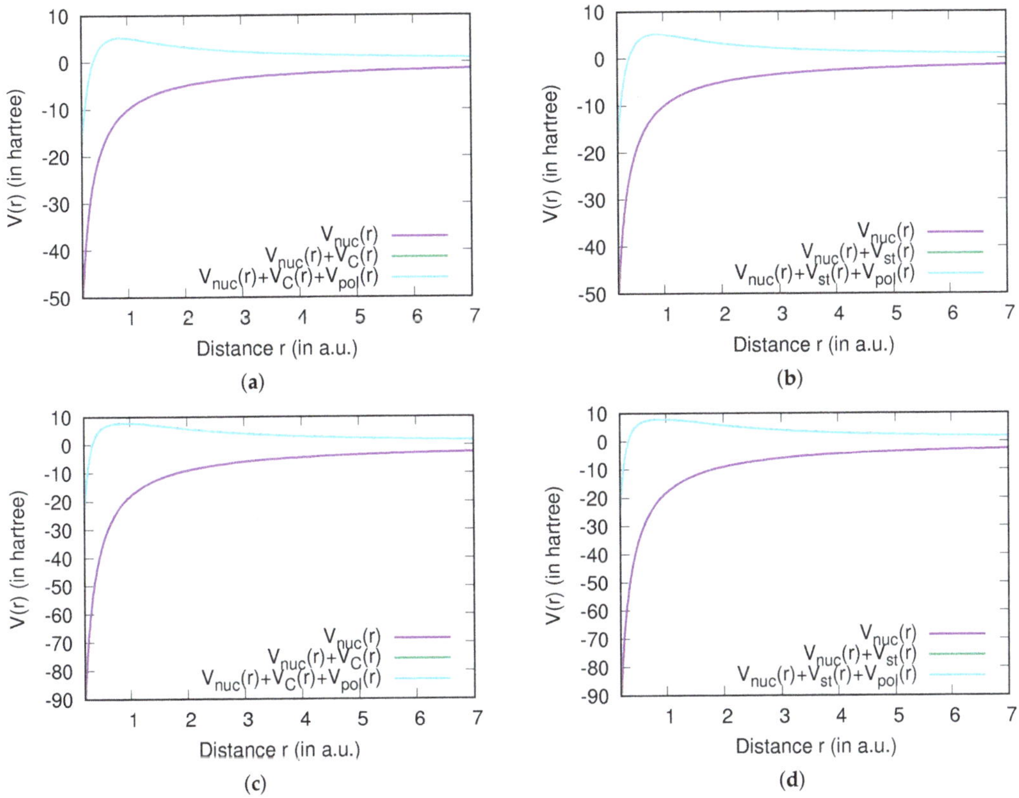

Figure 4. Plots demonstrating different contributions to $V(r)$ from the Ne and Ar noble atoms. In (**a**,**b**), results are given from the RCCSD and RNCCSD methods, respectively, for the Ne atom, while the results from the RCCSD and RNCCSD methods for the Ar atom are shown in (**c**,**d**), respectively. All quantities are given in a.u.

5. Conclusions

We have demonstrated approaches employing relativistic-coupled cluster theory to determine potentials for the evaluation of electron-atom elastic scattering cross-sections. For this purpose, we considered both the standard and normal versions of the relativistic coupled-cluster theory in the singles and doubles approximation, and presented results for the Be, Mg, Ne and Ar atoms as representative elements for the alkaline-earth and noble gas atoms of the Periodic Table. To estimate the static potential contributions, the finite-size nuclear effect was determined through the nuclear potential, while the two-electron correlation effects were estimated using relativistic coupled-cluster theory. The density functions of the above atoms from both the considered relativistic coupled-cluster theories were presented to estimate the Coulomb exchange potential contributions, which we neglected here for estimating potentials. Furthermore, we determined the electric dipole, quadrupole and dipole-quadrupole polarizabilities to account for the electron polarization effects on the scattering potential. The results from both the standard and normal relativistic coupled-cluster theories were compared with the literature values. These methods can be further applied to other heavier atomic systems to study electron-atom scattering cross-sections more accurately where the electron correlation effects within the atom will be more prominent than for the lighter elements investigated here.

Funding: This research received no external funding.

Acknowledgments: We acknowledge use of the Vikram-100 HPC cluster of the Physical Research Laboratory (PRL), Ahmedabad, India for the computations.

Conflicts of Interest: The author declares no conflict of interest.

References

1. Johnson, R.E. *Introduction to Atomic and Molecular Collisions*; Plenum Press: New York, NY, USA; London, UK, 1982.
2. Amusia, M.Y. Many-body atomic physics, Chapter 8. In *Many-Body Effects in Single Photoionization Processes*; Boyle, J.J., Pindzola, M.S., Eds.; Cambridge University Press: New York, NY, USA, 1998; p. 185.
3. Kotlarchyk, M. *Encyclopedia of Spectroscopy and Spectrometry*; Lindon, J.C., Ed.; Elsevier: Oxford, UK, 1999; pp. 2074–2084.
4. Pradhan, A.K.; Nahar, S.N. *Atomic Astrophysics and Spectroscopy*; Cambridge University Press: New York, NY, USA, 2011.
5. Dressler, R.A.; Chiu, Y.-H.; Zatsarinny, O.; Bartschat, K.; Srivastava, R.; Sharma, L. Near-infrared collisional radiative model for Xe plasma electrostatic thrusters: The role of metastable atoms. *J. Phys. D* **2009**, *42*, 185203. [CrossRef]
6. Dipti; Gangwar, R.K.; Srivastava, R.; Stauffer, A.D. Collisional-radiative model for non-Maxwellian inductively coupled argon plasmas using detailed fine-structure relativistic distorted-wave cross sections. *Eur. Phys. J. D* **2013**, *67*, 40244.
7. Bray, I.; Fursa, D.V.; Kheifets, A.S.; Stelbovics, A.T. Electrons and photons colliding with atoms: Development and application of the convergent close-coupling method. *J. Phys. B* **2002**, *35*, R117. [CrossRef]
8. Burke, P.G. *R-Matrix Theory of Atomic Collisions*; Springer: Berlin/Heidelberg, Germany, 2013.
9. Post, D.E. A review of recent developments in atomic processes for divertors and edge plasmas. *J. Nucl. Mater.* **1995**, *220*, 143–157. [CrossRef]
10. Jablonski, A.; Salvat, F.; Powell, C.J. Comparison of Electron Elastic-Scattering Cross Sections Calculated from Two Commonly Used Atomic Potentials. *J. Phys. Chem. Ref. Data* **2004**, *33*, 409–451. [CrossRef]
11. O'Connell, J.K.; Lane, N.F. Nonadjustable exchange-correlation model for electron scattering from closed-shell atoms and molecules. *Phys. Rev. A* **1983**, *27*, 1893. [CrossRef]
12. Yuan, J.; Zhang, Z. The low-lying shape resonances in low-energy electron scattering with Be, Mg and Ca atoms. *J. Phys. B At. Mol. Opt. Phys.* **1989**, *22*, 2751. [CrossRef]
13. Franz, J. Positron-electron correlation-polarization potentials for the calculation of positron collisions with atoms and molecules. *Eur. Phys. J. D* **2017**, *71*, 44. [CrossRef]
14. Tenfen, W.; Barp, M.V.; Arretche, F. Low-energy elastic scattering of positrons by O_2. *Phys. Rev. A* **2019**, *99*, 022703. [CrossRef]
15. Arretche, F.; Andermann, A.M.; Seidel, E.P.; Tenfen, W.; Sahoo, B.K. Polarization effects, shape resonances and bound states in low energy positron elastic scattering by Zinc and Cadmium vapours. *J. Electron Spectrosc. Relat. Phenom* **2022**, *254*, 147186. [CrossRef]
16. Khandker, M.H.; Haque, A.K.F.; Haque, M.M.; Billah, M.M.; Watabe, H.; Uddin, M.A. Relativistic Study on the Scattering of $e^{\pm}$ from Atoms and Ions of the Rn Isonuclear Series. *Atoms* **2021**, *9*, 59. [CrossRef]
17. Miller, W.H. Distorted-Wave Theory for Collisions of an Atom and a Diatomic Molecule. *J. Chem. Phys.* **1968**, *49*, 2373–2381. [CrossRef]
18. Madison, D.H.; Shelton, W.N. Distorted-Wave Approximation and Its Application to the Differential and Integrated Cross Sections for Electron-Impact Excitation of the $2\,^1P$ State of Helium. *Phys. Rev. A* **1973**, *7*, 499. [CrossRef]
19. Toshima, N.; Eichler, J. Distorted-wave approximations for relativistic atomic collisions. *Phys. Rev. A* **1990**, *41*, 5221. [CrossRef]
20. Srivastava, R.; Muktavat, K.; Stauffer, A.D. Electron Impact Excitation of Atoms in the Relativistic Distorted Wave Approximation. *Phys. Scr.* **2004**, *2004*, 241. [CrossRef]
21. Sharma, L.; Surzhykov, A.; Srivastava, R.; Fritzsche, S. Electron-impact excitation of singly charged metal ions. *Phys. Rev. A* **2011**, *83*, 062701. [CrossRef]
22. Marucha, A.M.; Kariuki, P.K.; Okumu, J.; Singh, C.S. Relativistic distorted wave approach to electron impact excitation of argon gas using a complex potential. *J. Phys. Commun.* **2021**, *5*, 075011. [CrossRef]
23. Teubner, P.J.O.; Farrell, P.M.; Karaganov, V.; Law, M.R.; Suvorov, V. Laser Assisted Collisions of Electrons with Metal Vapours. *Aust. J. Phys.* **1996**, *49*, 481–498. [CrossRef]
24. Joshipura, K.; Mason, N. *Atomic-Molecular Ionization by Electron Scattering: Theory and Applications*; Cambridge University Press: New York, NY, USA, 2019; pp. 177–218.
25. Buckman, S.J.; Cooper, J.W.; Elford, M.T.; Inokuti, M.; Itikawa, Y.; Tawara, H. *Interactions of Photons and Electrons with Atoms*; Itikawa, Y., Ed.; Springer: Berlin/Heidelberg, Germany, 2000; Volume 17A, pp. 1–77.
26. Jhanwar, B.L.; Khare, S.P.; Kumar, A., Jr. Elastic scattering of electrons on Ne atoms at intermediate energies. *J. Phys. B At. Mol. Phys.* **1978**, *11*, 887. [CrossRef]
27. Khare, S.P.; Kumar, A., Jr. Elastic scattering of electrons by argon atoms. *Pramana* **1978**, *10*, 63–73. [CrossRef]
28. Khare, S.P.; Kumar, A.; Lata, K. Elastic scattering of electrons and positrons by magnesium atoms at intermediate energies. *J. Phys. B At. Mol. Phys.* **1983**, *16*, 4419. [CrossRef]
29. Elkilany, S.A. Elastic scattering of electrons by helium and neon atoms. *Nouv. Cim. D* **1998**, *20*, 147–154. [CrossRef]
30. Phelps, A.V.; Greene, C.H.; Burke, J.P., Jr.; Phelps, A.V. Collision cross sections for argon atoms with argon atoms for energies from 0.01 eV to 10 keV. *J. Phys. B At. Mol. Opt. Phys.* **2000**, *33*, 2965. [CrossRef]

31. Khoperskiĭ, A.N.; Yavna, V.A.; Nadolinskiĭ, A.M.; Dzyuba, D.V. Elastic scattering of a photon by the neon atom near the K ionization edge. *Opt. Spectrosc.* **2004**, *96*, 195–197. [CrossRef]
32. Baynard, T.; Reber, A.C.; Niedziela, R.F.; Darveau, S.A.; Prutzman, B.; Berry, R.S. Electron-Atom Superelastic Scattering in Magnesium at Millielectron Volt Energies. *J. Phys. Chem. A* **2007**, *111*, 12487–12494. [CrossRef] [PubMed]
33. Kumar, A.; Kumar, S.; Rastogi, N.; Raj, D. Critical points for electron-Mg atom elastic scattering. *J. Phys. B At. Mol. Opt. Phys.* **2018**, *51*, 035203. [CrossRef]
34. Bishop, R.; Arponen, J.; Pajanee, P. *Aspects of Many-Body Effects in Molecules and Extended Systems*; Springer: Berlin/Heidelberg, Germany, 1989.
35. Szabo, A.; Ostuland, N. *Modern Quantum Chemistry*, 1st ed. (revised); Dover Publications, Inc.: Mineola, NY, USA, 1996.
36. Crawford, T.D.; Schaefer, H.F., III. *An Introduction to Coupled Cluster Theory for Computational Chemists*; Lipkowitz, K.B., Boyd, D.B., Eds.; Wiley: New York, NY, USA, 2000; pp. 33–136.
37. Shavitt, I.; Bartlett, R.J. *Many-Body Methods in Chemistry and Physics*; Cambidge University Press: Cambridge, UK, 2009.
38. Sharma, L.; Sahoo, B.K.; Malkar, P.; Srivastava, R. Application of relativistic coupled-cluster theory to electron impact excitations of Mg^+ in the plasma environment. *Eur. Phys. J. D* **2018**, *72*, 10. [CrossRef]
39. Bharti, S.; Sharma, L.; Sahoo, B.K.; Malkar, P.; Srivastava, R. Application of relativistic coupled cluster theory to elastic scattering of electrons from confined Ca atoms. *J. Phys. B* **2019**, *52*, 185003. [CrossRef]
40. Gribakin, G.F.; Ludlow, J. Convergence of partial-wave expansions for energies, scattering amplitudes and positron annihilation rates. *J. Phys. B At. Mol. Opt. Phys.* **2002**, *35*, 339. [CrossRef]
41. Chakraborty, A.; Rithvik, S.K.; Sahoo, B.K. Relativistic normal coupled-cluster theory analysis of second- and third-order electric polarizabilities of Zn I. *Phys. Rev. A* **2022**, *105*, 062815. [CrossRef]
42. Newton, R.G. *Scattering Theory of Waves and Particles*; Springer: Berlin/Heidelberg, Germany, 1982.
43. Estevez, G.; Bhuiyan, L.B. Electrostatic potential due to a Fermi-type charge distribution. *Am. J. Phys.* **1985**, *53*, 450–453. [CrossRef]
44. Hara, S. The Scattering of Slow Electrons by Hydrogen Molecules. *J. Phys. Soc. Jpn.* **1967**, *22*, 710–718. [CrossRef]
45. Arretche, F.; Barp, M.V.; Tenfen, W.; Seidel, E.P. The Hidden Ramsauer-Townsend Effect in Positron Scattering by Rare Gas Atoms. *Braz. J. Phys.* **2020**, *50*, 844–856. [CrossRef]
46. Burrow, P.; Michejda, J.; Comer, J. Low-energy electron scattering from Mg, Zn, Cd and Hg: Shape resonances and electron affinities. *J. Phys. B At. Mol. Phys.* **1976**, *9*, 3225. [CrossRef]
47. Sahoo, B.K.; Das, B.P. Relativistic coupled-cluster studies of dipole polarizabilities in closed-shell atoms. *Phys. Rev. A* **2008**, *77*, 062516. [CrossRef]
48. Singh, Y.; Sahoo, B.K. Correlation trends in the polarizabilities of atoms and ions in the boron, carbon, and zinc homologous sequences of elements. *Phys. Rev. A* **2014**, *90*, 022511. [CrossRef]
49. Singh, Y.; Sahoo, B.K. Rigorous limits on the hadronic and semileptonic CP-violating coupling constants from the electric dipole moment of ^{199}Hg. *Phys. Rev. A* **2015**, *91*, 030501(R). [CrossRef]
50. Prasannaa, V.S.; Mitra, R.; Sahoo, B.K. Reappraisal of P, T-odd parameters from the improved calculation of electric dipole moment of ^{225}Ra atom. *J. Phys. B* **2020**, *53*, 195004. [CrossRef]
51. Sahoo, B.K.; Das, B.P. Relativistic Normal Coupled-Cluster Theory for Accurate Determination of Electric Dipole Moments of Atoms: First Application to the ^{199}Hg Atom. *Phys. Rev. Lett.* **2018**, *120*, 203001. [CrossRef]
52. Corzo, H.H.; Sehanobish, A.; Kara, O. Learning Full Configuration Interaction Electron Correlations with Deep Learning. *arXiv* **2021**, arXiv:2106.08138.
53. Zaklika, J.; Hładyszowski, J.; Ordon, P.; Komorowski, L. From the Electron Density Gradient to the Quantitative Reactivity Indicators: Local Softness and the Fukui Function. *ACS Omega* **2022**, *7*, 7745–7758. [CrossRef] [PubMed]
54. Diouf, Y.; Talla, K.; Diallo, S.; Gomis, L. Numerical Study of Density Functional Theory of Multi-electronic Atoms: Case of Carbon and Helium. *Am. J. Nanomaterials* **2021**, *9*, 12–22. [CrossRef]
55. Thakkar, A.J.; Lupinetti, C. Atomic Polarizabilities and Hyperpolarizabilities: A Critical Compilation. In *Computational, Numerical and Mathematical Methods in Sciences and Engineering*; Maroulis, G., Ed.; 2006; Volume 1, Chapter 4, pp. 505–529. Available online: https://www.worldscientific.com/doi/abs/10.1142/9781860948862_0014 (accessed on 24 July 2022).
56. Porsev, S.G.; Derevianko, A. High-accuracy calculations of dipole, quadrupole, and octupole electric dynamic polarizabilities and van der Waals coefficients C_6, C_8, and C_{10} for alkaline-earth dimers. *J. Expt. Theor. Phys.* **2006**, *102*, 195–205. [CrossRef]
57. Singh, Y.; Sahoo, B.K.; Das, B.P. Correlation trends in the ground-state static electric dipole polarizabilities of closed-shell atoms and ions. *Phys. Rev. A* **2013**, *88*, 062504. [CrossRef]
58. Schwerdtfeger, P.; Nagle, J.K. 2018 Table of static dipole polarizabilities of the neutral elements in the periodic table. *Mol. Phys.* **2019**, *117*, 1200–1225. [CrossRef]
59. Huot, J.; Bose, T.K. Experimental determination of the dielectric virial coefficients of atomic gases as a function of temperature. *J. Chem. Phys.* **1991**, *95*, 2683–2687. [CrossRef]
60. Taylor, P.R.; Lee, T.J.; Rice, J.E.; Almlöf, J. The polarizabilities of neon. *Chem. Phys. Lett.* **1989**, *163*, 359–365. [CrossRef]
61. Johnston, D.R.; Oudemans, G.J.; Cole, R.H. Dielectric Constants of Imperfect Gases. I. Helium, Argon, Nitrogen, and Methane. *J. Chem. Phys.* **1960**, *33*, 1310–1317. [CrossRef]
62. Hohm, U.; Kerl, K. Interferometric measurements of the dipole polarizability α of molecules between 300 K and 1100 K. *Mol. Phys.* **1990**, *69*, 803–817. [CrossRef]

63. Cernusak, I.; Diercksen, G.H.F.; Sadlej, A.J. Multipole polarizabilities of Ar. *Chem. Phys. Letts.* **1986**, *128*, 18–24. [CrossRef]
64. Maroulis, G.; Bishop, D.M. On the electric polarisabilities of argon. *J. Phys. B At. Mol. Phys.* **1985**, *18*, 4675. [CrossRef]

atoms

MDPI

Article

Magnetic Sublevel Independent Magic and Tune-Out Wavelengths of the Alkaline-Earth Ions

Jyoti [1], Harpreet Kaur [1], Bindiya Arora [1,*] and Bijaya Kumar Sahoo [2]

1 Department of Physics, Guru Nanak Dev University, Amritsar 143005, Punjab, India; jyotiphy.rsh@gndu.ac.in (J.); harpreetphy.rsh@gndu.ac.in (H.K.)
2 Physical Research Laboratory, Atomic, Molecular and Optical Physics Division, Navrangpura, Ahmedabad 380009, Gujarat, India; bijaya@prl.res.in
* Correspondence: bindiya.phy@gndu.ac.in

Abstract: Light shift in a state due to the applied laser in an atomic system vanishes at tune-out wavelengths (λ_Ts). Similarly, differential light shift in a transition vanishes at the magic wavelengths (λ_{magic}s). In many of the earlier studies, values of the electric dipole (E1) matrix elements were inferred precisely by combining measurements and calculations of λ_{magic}. Similarly, the λ_T values of an atomic state can be used to infer the E1 matrix element, as it involves dynamic electric dipole (α) values of only one state whereas the λ_{magic} values require evaluation of α values for two states. However, both the λ_{magic} and λ_T values depend on angular momenta and their magnetic components (M) of states. Here, we report the λ_{magic} and λ_T values of many $S_{1/2}$ and $D_{3/2,5/2}$ states, and transitions among these states of the Mg^+, Ca^+, Sr^+ and Ba^+ ions that are independent of M values. It is possible to infer a large number of E1 matrix elements of the above ions accurately by measuring these values and combining with our calculations.

Keywords: relativistic all-order method; magic wavelength; tune-out wavelength; dipole polarizability

Citation: Jyoti; Kaur, H.; Arora, B.; Sahoo, B.K. Magnetic Sublevel Independent Magic and Tune-Out Wavelengths of the Alkaline-Earth Ions. *Atoms* **2022**, *10*, 72. https://doi.org/10.3390/atoms10030072

Academic Editor: Jean-Paul Mosnier

Received: 10 June 2022
Accepted: 6 July 2022
Published: 11 July 2022

Publisher's Note: MDPI stays neutral with regard to jurisdictional claims in published maps and institutional affiliations.

1. Introduction

Singly charged alkaline-earth ions are the most eligible candidates for considering for high-precision measurements due to several advantages [1]. Except Be^+ and Mg^+, other alkaline-earth ions have two metastable states and most of the transitions among the ground and metastable states are accessible by lasers. This is why these ions are considered for carrying out high-precision measurements such as testing Lorentz symmetry violations [2–4], parity nonconservation effects [5], non-linear isotope shift effects [6], quantum information [7,8] and many more, including for the optical atomic clock experiments [9]. Above all, optical lattices blended with unique features of optical transitions in the ionic system lead to the revolution in the clock frequency states. Since the confinement of Mg^+ ions in a monochromatic optical dipole trap has become feasible experimentally for several ms [10], the pathways to implement these ions to realize optical lattice clocks have been opened up due to the fact that ions provide more accurate atomic clocks since various systematics in the ions can be controlled easily [11,12].

One of the major systematics in the aforementioned high-precision measurements is the Stark shift due to the employed laser, which depends on the frequency of the laser. The solution to this problem was suggested by Katori et al. [13], who proposed that the trapping laser can be tuned to wavelengths at which differential ac Stark shifts of the transitions can vanish [13]. These wavelengths were coined as magic wavelengths (λ_{magic}s) and are being popularly used in optical lattice clocks. The development of the technique to measure λ_{magic}s in the singly charged $^{40}Ca^+$ ion [14] has opened up the possibilities to measure these wavelengths in the other ions, such as Al^+, Hg^+, Yb^+, Cd^+, In^+ and so on, that are being considered in many high-precision measurements. Further, the infrared magic wavelength has been identified recently for all optical trapping of $^{40}Ca^+$ ion clock [15]. The stability of optical lattices can be enhanced by optical cavities in various dimensions while availing the coordinated choice of λ_{magic}s [16] and state-dependent potentials [17], which can further be implemented in the all-optical confinement of these ions. In quantum state engineering [18], λ_{magic}s provide an opportunity to extract accurate values of oscillator strengths [19], which are particularly important for correct stellar modeling and analysis of spectral lines identified in the spectra of stars and other heavenly bodies so as to infer fundamental stellar parameters [20,21].

Apart from the magic trapping condition, where light shift in two internal states is identical, another well-known limiting case is where light shift in one state vanishes. This case is known as a tune-out condition [22]. Applications of such tune-out wavelengths (λ_T) lie in novel cooling techniques of atoms [23], selective addressing and manipulation of quantum states [24–26], precision measurement of atomic structures [27–32] and precise estimation of oscillator strength ratios [33]. Additionally, tune-out conditions are powerful tools for the evaporative cooling of optical lattices [34] and, hence, are important for experimental explorations.

In one of the experiments pertaining to magic wavelengths of alkaline-earth ions, Liu et al. demonstrated the existence of magic wavelengths for a single trapped $^{40}Ca^+$ ion [14], whereas Jiang et al. evaluated magic wavelengths of Ca^+ ions for linearly and circularly polarized light using the relativistic configuration interaction plus core polarization (RCICP) approach [35,36]. Recently, Chanu et al. proposed a model to trap Ba^+ ions by inducing an ac Stark shift using a 653 nm linearly polarized laser [37]. Kaur et al. reported magic wavelengths for $nS_{1/2} - nP_{1/2,3/2}$ and $nS_{1/2} - mD_{3/2,5/2}$ transitions in alkaline-earth-metal ions using linearly polarized light [16], whereas Jiang et al. located magic and tune-out wavelengths for Ba^+ ions using the RCICP approach [38]. Despite having a large number of applications, these magic wavelengths suffer a setback because of their dependency on the magnetic-sublevels (M) of the atomic systems. Linearly polarized light has been widely used for the trapping of atoms and ions as it is free from the contribution of the vector component in the interaction between atomic states and electric fields. However, the magic wavelengths thus identified are again magnetic-sublevel dependent for the transitions involving states with angular momenta greater than 1/2. On the other hand, the cooling of ions using circularly polarized light requires magnetic-sublevel selections. In order to circumvent this M dependency of magic wavelengths, a magnetic-sublevel-independent strategy for the trapping of atoms and ions was proposed by Sukhjit et al. [39]. Later on, Kaur et al. implemented a similar technique to compute magic and tune-out wavelengths independent of magnetic sublevels M for different $nS_{1/2} - (n-1)D_{3/2,5,2}$ transitions in Ca^+, Sr^+ and Ba^+ ions corresponding to n = 4 for Ca^+, 5 for Sr^+ and 6 for Ba^+ ion [40].

In addition to the applications of λ_{magic} in getting rid of differential Stark shift in a transition, they are also being used to infer the electric dipole (E1) matrix elements of many allowed transitions in different atomic systems [14,38,41,42]. The basic procedure of these studies is that the λ_{magic} values are calculated by fine-tuning of the magnitudes of dominantly contributing E1 matrix elements to reproduce their measured values. Then, the set of the E1 matrix elements that gives rise to the best matched λ_{magic} values are considered as the recommended E1 matrix elements. However, calculations of these λ_{magic} values of a transition demand the determination of the dynamic E1 polarizabilities (α) of both the states. In view of this, use of the λ_T values of a given atomic state can be advantageous, as they involve dynamic α values of only one state. Furthermore, both the λ_{magic} and λ_T values depend on angular momenta and their magnetic components (M) of atomic states. This requires evaluation of scalar, vector and tensor components of the α values for states with angular momenta greater than 1/2, which is very cumbersome. To circumvent this problem, we present here M-sublevel-independent λ_{magic} and λ_T values of many states and transitions involving a number of $S_{1/2}$ and $D_{3/2,5/2}$ states in the alkaline-earth metal ions from Mg^+ through Ba^+ that can be inferred to the E1 matrix elements more precisely. The basic idea behind this work is to use the calculated E1 matrix elements to predict the M-independent λ_{magic} and λ_T values of the above ions. Theoretical determinations of these λ_{magic} and λ_T are very sensitive to the magnitudes of a few principal E1 matrix elements. When measurements of these quantities are available, by combining them with our calculations it is possible to fine tune the E1 matrix element values to higher precision. At this stage, our reported λ_{magic} and λ_T values will be useful to guide the experimentalists to carry out these measurements. The measurements can be achieved by setting up experiments suitably by fixing the polarization and quantization angles of the applied lasers. To validate our results for the transitions involving high-lying states, we compared the values of our λ_{magic} and λ_T values for the ground to the metastable states of the considered alkaline-earth ions with the previously reported values.

The paper is organized as follows: In Section 2, we provide underlying theory and Section 3 describes the method of evaluation of the calculated quantities. Section 4 discusses the obtained results, while concluding the study in Section 5. Unless we have stated explicitly, physical quantities are given in atomic units (a.u.).

2. Theory

The electric field $\mathcal{E}(r,t)$ associated with a general plane electromagnetic wave can be represented in terms of complex polarization vector $\hat{\chi}$ and the real wave vector **k** by the following expression [43]

$$\mathcal{E}(\mathbf{r},t) = \frac{1}{2}\mathcal{E}\hat{\chi}e^{-\imath(\omega t-\mathbf{k}.\mathbf{r})} + c.c., \tag{1}$$

where $c.c.$ is the complex conjugate of the preceding term. Assuming $\hat{\chi}$ to be real and adopting the coordinate system as presented in Figure 1, the polarization vector can be expressed as [39]

$$\hat{\chi} = e^{\imath\sigma}(cos\phi\ \hat{\chi}_{maj} + \imath\ sin\phi\ \hat{\chi}_{min}), \tag{2}$$

where $\hat{\chi}_{maj}$ and $\hat{\chi}_{min}$ denote the real components of the polarization vector $\hat{\chi}$, σ is the real quantity denoting the arbitrary phase and ϕ is analogous to degree of polarization A such that $A = sin(2\phi)$. For linearly polarized light, $\phi = 0$ whereas ϕ takes the value either $\pi/4$ or $3\pi/4$ for circularly polarized light, which further defines $A = 0$ for linearly polarized and $A = 1(-1)$ for right-hand (left-hand) circularly polarized light [43]. As shown in Figure 1, this coordinate system follows

$$cos^2\theta_p = cos^2\phi\ cos^2\theta_{maj} + sin^2\phi\ sin^2\theta_{min} \tag{3}$$

and

$$\theta_{maj} + \theta_{min} = \frac{\pi}{2}. \tag{4}$$

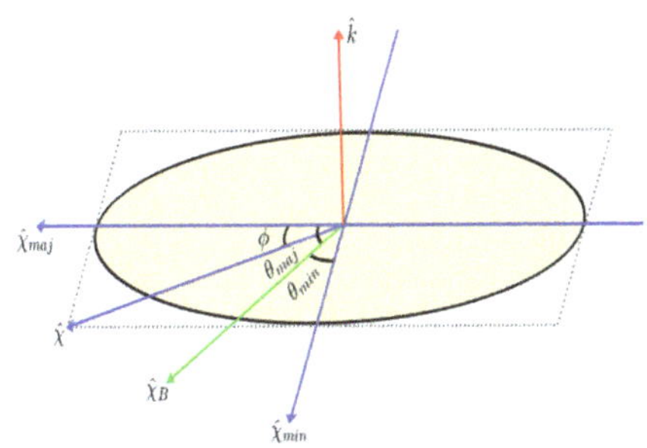

Figure 1. Representation of elliptically polarized laser beam swept out by the laser's polarization vector in one period. $\hat{\chi}$ represents the laser's complex polarization vector and $\hat{k}$ as the laser wave vector perpendicular to quantization axis $\hat{\chi}_B$. The vectors $\hat{\chi}_{maj}$, $\hat{\chi}_{min}$ and $\hat{k}$ are mutually perpendicular to each other.

Here, θ_p is the angle between quantization axis $\hat{\chi}_B$ and direction of polarization vector $\hat{\chi}$ and the parameters θ_{maj} and θ_{min} are the angles between respective unit vectors and $\hat{\chi_B}$.

When an atomic system is subjected to the above electric field and the magnitude of $\mathcal{E}$ is small, a shift in the energy of its nth level (Stark shift) can be given by

$$\delta E_n^K \simeq -\frac{1}{2}\alpha_n^K(\omega)|\mathcal{E}|^2, \tag{5}$$

where $\alpha_n^K(\omega)$ is known as the second-order electric dipole (E1) polarizability and the superscript K denotes angular momentum of the state, which can be atomic angular momentum J or hyperfine-level angular momentum F. Depending upon polarization, dynamic dipole polarizability $\alpha_n^K(\omega)$ can be expressed as

$$\begin{aligned}\alpha_n^K(\omega) &= \alpha_{nS}^K(\omega) + \beta(\chi)\tfrac{M_K}{2K}\alpha_{nV}^K(\omega) \\ &+\gamma(\chi)\frac{3M_K^2-K(K+1)}{K(2K-1)}\alpha_{nT}^K(\omega),\end{aligned} \tag{6}$$

where α_{nS}^K, α_{nV}^K and α_{nT}^K are the scalar, vector and tensor components of the polarizability, respectively. The expression can be defined on the basis of the coordinate system provided in the Figure 1. Geometrically, values for $\beta(\chi)$ and $\gamma(\chi)$ in their elliptical form are given as [39,43]

$$\beta(\chi) = \iota(\hat{\chi}\times\hat{\chi}^*).\hat{\chi}_B = Acos\ \theta_k \tag{7}$$

and

$$\gamma(\chi) = \frac{1}{2}[3(\hat{\chi}^*.\hat{\chi}_B)(\hat{\chi}.\hat{\chi}_B) - 1] = \frac{1}{2}\left(3cos^2\ \theta_p - 1\right), \tag{8}$$

where θ_k is the angle between direction of propagation $\mathbf{k}$ and $\hat{\chi}_B$. Substitution of $\beta(\chi)$ and $\gamma(\chi)$ from Equations (7) and (8) reforms the expression for dipole polarizability to

$$\begin{aligned}\alpha_n^K(\omega) &= \alpha_{nS}^K(\omega) + Acos\theta_k\tfrac{M_K}{2K}\alpha_{nV}^K(\omega) \\ &+\left(\frac{3cos^2\theta_p-1}{2}\right)\frac{3M_K^2-K(K+1)}{K(2K-1)}\alpha_{nT}^K(\omega)\end{aligned} \tag{9}$$

with the azimuthal quantum number M_K of the respective angular momentum K.

Thus, it is obvious from Equation (6) that the α_n^K values of two states have to be same if we intend to find λ_{magic} for the transition involving both the states. Since the above expression for α_n^K has M_K dependency, the λ_{magic} become M_K dependent. In order to remove M_K dependency, one can choose $M_K = 0$ sublevels but in the atomic states of the alkaline-earth ions they are non-zero while isotopes with integer nuclear spin of the alkaline-earth ions M_Ks are again non-zero. To address this, a suitable combination of the $\beta(\chi)$ and $\gamma(\chi)$ parameters needs to be chosen such that $cos\theta_k = 0$ and $cos^2\theta_p = \frac{1}{3}$, which are feasible to achieve in an experiment by setting θ_k, $\hat{\chi}_{maj}$ and ϕ values, as demonstrated in Ref. [39]. In such a scenario, the λ_{magic} values can depend on the scalar part only by suppressing the vector and tensor components of α_n^K; i.e., the net differential Stark effect of a transition occurring from the J to J' states will be given by

$$\delta E_{JJ'} = -\frac{1}{2}\left[\alpha_{nS}^J(\omega) - \alpha_{nS}^{J'}(\omega)\right]\mathcal{E}^2. \tag{10}$$

This has an additional advantage that the differential Stark effects at an arbitrary electric field become independent of the choice of atomic or hyperfine levels in a given atomic system, as the scalar component of α_n^J and α_n^F are the same. Again, the same choice of λ_{magic} values will be applicable to both the atomic and hyperfine levels in a high-precision experiment.

3. Method of Evaluation

Determination of α_n^J values requires accurate calculations of E1 matrix elements. For the computation of E1 matrix elements, we need accurate atomic wave functions of the alkaline-earth ions. We employed here a relativistic all-order method to determine the atomic wave functions of the considered atomic systems, whose atomic states have a closed-core configuration with an unpaired electron in the valence orbital. Detailed descriptions of our all-order method can be found in Refs. [44–47]; however, a brief outline of the same is also provided here for the completeness.

Our all-order method follows the relativistic coupled-cluster (RCC) theory ansätz

$$|\psi_v\rangle = e^S|\phi_v\rangle, \tag{11}$$

where $|\phi_v\rangle$ represents the mean-field wave function of the state v and constructed as [48]

$$|\phi_v\rangle = a_v^\dagger|0_c\rangle, \tag{12}$$

where $|0_c\rangle$ represents the Dirac–Hartree–Fock (DHF) wave function of the closed core. Subscript v represents the valence orbital of the considered state. In our calculations, we consider only linear terms in the singles and doubles approximation of the RCC theory (SD method) by expressing [48]

$$|\psi_v\rangle = (1 + S_1 + S_2 + ...)|\phi_v\rangle, \tag{13}$$

where S_1 and S_2 depict terms corresponding to the single and double excitations, respectively, that can further be written in terms of second quantization creation and annihilation operators as follows [49]

$$S_1 = \sum_{ma} \rho_{ma} a_m^\dagger a_a + \sum_{m \neq v} \rho_{mv} a_m^\dagger a_v \tag{14}$$

and

$$S_2 = \frac{1}{2} \sum_{mnab} \rho_{mnab} a_m^\dagger a_n^\dagger a_b a_a + \sum_{mna} \rho_{mnva} a_m^\dagger a_n^\dagger a_a a_v, \tag{15}$$

where indices m and n range over all possible virtual orbitals, and indices a and b range over all occupied core orbitals. The coefficients ρ_{ma} and ρ_{mv} represent excitation coefficients of the respective single excitations for the core and the valence electrons, respectively, whereas ρ_{mnab} and ρ_{mnva} depict double excitation coefficients for the core and the valence electrons, respectively. These amplitudes are calculated in an iterative procedure [50], due to which they include electron correlation effects to all-order.

Hence, atomic wave function of the considered states in the alkaline-earth ions are expressed as [48,51]:

$$\begin{aligned} |\psi_v\rangle_{SD} &= \Big[1 + \textstyle\sum_{ma} \rho_{ma} a_m^\dagger a_a + \frac{1}{2} \sum_{mnab} \rho_{mnab} a_m^\dagger a_n^\dagger a_b a_a \\ &\quad + \textstyle\sum_{m \neq v} \rho_{mv} a_m^\dagger a_v + \sum_{mna} \rho_{mnva} a_m^\dagger a_n^\dagger a_a a_v\Big]|\phi_v\rangle. \end{aligned} \tag{16}$$

To improve the calculations further and understand the importance of contributions from the triple excitations in the RCC theory, we take into account important core and valence triple excitations through the perturbative approach over the SD method (SDpT method) by redefining the wave function expression as [48]

$$\begin{aligned} |\psi_v\rangle_{SDpT} &= |\psi_v\rangle_{SD} + \Big[\tfrac{1}{6} \textstyle\sum_{mnrab} \rho_{mnrvab} a_m^\dagger a_n^\dagger a_r^\dagger a_b a_a a_v \\ &\quad + \tfrac{1}{18} \textstyle\sum_{mnrabc} \rho_{mnrabc} a_m^\dagger a_n^\dagger a_r^\dagger a_c a_b a_a\Big]|\phi_v\rangle. \end{aligned} \tag{17}$$

After obtaining the wave functions of the interested atomic states, we evaluate the E1 matrix elements between states $|\psi_v\rangle$ and $|\psi_w\rangle$ as [49]

$$D_{wv} = \frac{\langle\psi_w|D|\psi_v\rangle}{\sqrt{\langle\psi_w|\psi_w\rangle\langle\psi_v|\psi_v\rangle}}, \tag{18}$$

where $D = -e\Sigma_j \mathbf{r}_j$ is the E1 operator with $\mathbf{r}_j$ being the position of jth electron [51]. The resulting expression of numerator of Equation (18) includes the sum of the DHF matrix elements z_{wv}; twenty correlation terms of the SD method that are linear or quadratic functions of excitation coefficients ρ_{mv}, ρ_{ma}, ρ_{mnva} and ρ_{mnab}; and their core counterparts [45].

In the sum-over-states approach, expression for the scalar dipole polarizability is given by

$$\alpha_v(\omega) = \frac{2}{3(2J_v+1)} \sum_{v\neq w} \frac{(E_v - E_w)|\langle\psi_v||D||\psi_w\rangle|^2}{(E_v - E_w)^2 - \omega^2}, \tag{19}$$

where $\langle\psi_v||D||\psi_w\rangle$ is the reduced matrix element for the transition occurring between the states involving the valence orbitals v and w. Here, we dropped the superscript J in the dipole polarizability notation for the brevity. For convenience, we divide the entire contribution to $\alpha_n(\omega)$ for any state v, into three parts, as

$$\alpha_v = \alpha_{v,c} + \alpha_{v,vc} + \alpha_{v,val} \tag{20}$$

where c, vc and val corresponds to core, valence-core and valence contributions arising due to the correlations among the core orbitals, core-valence orbitals and valence-virtual orbitals, respectively [52]. Due to much smaller magnitudes, the core and core-valence contributions are calculated using the DHF method. The dominant contributions will arise valence from $\alpha_{v,val}$ due to small energy denominators. Again, the high-lying states will not contribute to $\alpha_{v,val}$ owing to large energy denominators. Thus, we calculate E1 matrix elements only among the low-lying excited states and refer the contributions as 'Main'. Contributions from the less contributing high-lying states are referred as 'Tail' and are estimated again using the DHF method. To reduce the uncertainties in the estimations of Main contributions, we used experimental energies of the states from the National Institute of Science and Technology atomic database (NIST AD) [53].

4. Results and Discussion

The precise computation of magic and tune-out wavelengths requires the accurate determination of E1 matrix elements as well as dipole polarizabilities. In our work, we used E1 matrix elements available on Portal for High-Precision Atomic Data and Computation [54] and NIST Atomic Spectra Database [53], respectively.

We listed resonance transitions, magic wavelengths and their corresponding polarizabilities along with their uncertainties for magnetic-sublevel independent $nS - mD$ transitions for alkaline-earth ions from Mg^+ through Ba^+ along with their comparison with the available literature in the Table 1 through Table 4, respectively. Results reported by Kaur et al. [40] were obtained by similar analyses, but contributions from some of the high-lying states were neglected in that work. Furthermore, we used excitation energies for most of the states listed in the NIST AD compared to Kaur et al. This resulted in more precise values of dipole polarizabilities, hence further improving the magic and tune-out wavelengths.

Table 1. Magic wavelengths λ_{magic} (in nm) with their corresponding polarizability $\alpha_n(\omega)$ (in a.u.) for $(3,4)S_{1/2} - 3D_{3/2,5/2}$ and $4S_{1/2} - 4D_{3/2}$ transitions in Mg^+ ion.

	$3S_{1/2} - 3D_{3/2}$				$3S_{1/2} - 3D_{5/2}$		
Resonance	λ_{res}	λ_{magic}	α_{magic}	**Resonance**	λ_{res}	λ_{magic}	α_{magic}
$3D_{3/2} \rightarrow 6P_{3/2}$	292.92			$3D_{5/2} \rightarrow 5F_{5/2}$	310.56		
		313.89 (2)	168.72 (13)			313.86 (2)	168.85 (13)
$3D_{3/2} \rightarrow 5P_{1/2}$	385.15			$3D_{5/2} \rightarrow 5P_{3/2}$	384.920		
		385.300 (1)	73.51 (7)			385.101 (1)	73.59 (6)
		757.8 (1.5)	40.43 (5)				
$3D_{3/2} \rightarrow 5P_{3/2}$	1091.83			$3D_{5/2} \rightarrow 4F_{5/2}$	448.24		
		1092.436 (1)	37.41 (5)			756.7 (1.5)	40.45 (5)
$3D_{3/2} \rightarrow 4P_{1/2}$	1095.48			$3D_{5/2} \rightarrow 4P_{3/2}$	1091.72		
	$4S_{1/2} - 3D_{3/2}$				$4S_{1/2} - 3D_{5/2}$		
$4S_{1/2} \rightarrow 5P_{3/2}$	361.48			$4S_{1/2} \rightarrow 5F_{7/2}$	310.56		
		361.26 (1)	−202.1 (1.3)			344.87 (47)	−160.00 (13)
$4S_{1/2} \rightarrow 5P_{1/2}$	361.66			$4S_{1/2} \rightarrow 5P_{3/2}$	361.48		
		361.627 (1)	−203.2 (1.3)			361.626 (1)	−203.8 (1.3)
$4S_{1/2} \rightarrow 4P_{3/2}$	922.08			$4S_{1/2} \rightarrow 5P_{1/2}$	361.66		
		923.812 (1)	−140.74 (97)				
$4S_{1/2} \rightarrow 4P_{1/2}$	924.68			$3D_{5/2} \rightarrow 5P_{3/2}$	384.93		
						385.408 (4)	−163.67 (13)
$3D_{3/2} \rightarrow 4P_{3/2}$	1091.83			$3D_{5/2} \rightarrow 4F_{5/2}$	448.24		
		1092.377 (1)	1976.9 (1.5)				
$3D_{3/2} \rightarrow 4P_{1/2}$	1095.48			$4S_{1/2} \rightarrow 4P_{3/2}$	922.08		
		1132.53 (6)	1681.4 (1.3)			923.812 (1)	−144.72 (98)
				$4S_{1/2} \rightarrow 4P_{1/2}$	924.68		
				$3D_{5/2} \rightarrow 4P_{3/2}$	1091.72		
						1128.42 (6)	1706.3 (1.3)
	$4S_{1/2} - 4D_{3/2}$				$4S_{1/2} - 4D_{5/2}$		
$4D_{3/2} \rightarrow 7F_{5/2}$	526.58			$4D_{5/2} \rightarrow 7F_{5/2,7/2}$	526.57		
		591.48 (1)	−422.01 (31)			591.34 (1)	−421.69 (31)
$4D_{3/2} \rightarrow 7P_{3/2}$	591.83			$4D_{5/2} \rightarrow 7P_{3/2}$	591.81		
		591.8603 (3)	−422.86 (31)				
$4D_{3/2} \rightarrow 7P_{1/2}$	591.98						
		616.11 (22)	−482.44 (35)			616.02 (23)	−482.19 (35)
$4D_{3/2} \rightarrow 6F_{5/2}$	634.87			$4D_{5/2} \rightarrow 6F_{7/2}$	634.85		
$4P_{1/2} \rightarrow 4D_{3/2}$	787.92						
		789.499 (4)	−1578.9 (1.2)				
$4P_{3/2} \rightarrow 4D_{3/2}$	789.82			$4P_{3/2} \rightarrow 4D_{5/2}$	789.85		
$4D_{3/2} \rightarrow 6P_{3/2}$	811.78			$4D_{5/2} \rightarrow 6P_{3/2}$	811.75		
		811.8186 (1)	−1973.6 (1.5)			812.118 (1)	−1980.0 (1.5)
						844.63 (10)	−2964.8 (2.2)
$4D_{3/2} \rightarrow 6P_{1/2}$	812.83						
		812.648 (1)	−1991.3 (1.5)				
		843.61 (11)	−2921.4 (2.2)				
$4S_{1/2} \rightarrow 4P_{3/2}$	922.08			$4S_{1/2} \rightarrow 4P_{3/2}$	922.08		
		923.839 (1)	−4957.9 (6.5)			923.839 (1)	−4975.8 (6.4)
$4S_{1/2} \rightarrow 4P_{1/2}$	924.68			$4S_{1/2} \rightarrow 4P_{1/2}$	924.68		
$4D_{3/2} \rightarrow 5F_{5/2}$	963.51			$4D_{5/2} \rightarrow 5F_{7/2}$	963.45		
		1006.10 (13)	3585.5 (2.7)			1005.86 (13)	3594.8 (2.7)

Further discussion regarding the magic wavelengths is provided in the Section 4.1 for the considered alkaline-earth ions. Furthermore, we discussed our results for tune-out wavelengths in the Section 4.2 along with the comparison of our results with respect to the available theoretical data.

4.1. Magic Wavelengths

4.1.1. Mg^+

In Table 1, we tabulated our results for magic wavelengths and their corresponding dipole polarizabilities for $(3,4)S_{1/2} - 3D_{3/2,5/2}$ and $4S_{1/2} - 4D_{3/2}$ transitions. Figure 2a demonstrates scalar dipole polarizabilities of $3S_{1/2}$ and $3D_{3/2,5/2}$ states of Mg^+ ion with respect to the wavelength of the external field. It can be perceived from the figure that a number of magic wavelengths at the crossings of the scalar polarizabilities' curves of the corresponding state have been predicted for the transition. As can be seen from Table 1, a total of 4 magic wavelengths were found for $3S - 3D_{3/2}$ transition, whereas the $3S - 3D_{5/2}$

transition shows a total of 3 magic wavelengths in the range 300–1250 nm, out of which no magic wavelength was found to exist in visible spectrum. However, all the magic wavelengths enlisted in Table 1 support red-detuned trap.

Figure 2b represents the plot of scalar dipole polarizabilities of 4*S* and $4D_{3/2,5/2}$ states against the wavelength of the external field. It can also be assessed from Table 1 that there exists a total of nine magic wavelengths in the considered wavelength range for $4S - 4D_{3/2}$ transition, whereas only five magic wavelengths are seen for the $4S - 4D_{5/2}$ transition. However, in both cases, all the magic wavelengths except those around 616 nm, 844 nm and 1006 nm are close to resonance, thereby making them unsuitable for further use. However, out of these three values, λ_{magic} at 616 nm lies in the visible region and is far-detuned with considerable deep potential. Hence, we recommend this magic wavelength for the trapping of Mg^+ ions for both $4S$–$4D_{3/2,5/2}$ transitions for further experimentations in optical clock applications.

Figure 2c demonstrates the magic wavelengths for the M_J-independent scheme for $4S$–$3D_{3/2,5/2}$ transitions for Mg^+ ions along with their corresponding scalar dynamic polarizabilities. According to Table 1, it can be realized that none of the magic wavelengths for these transitions lie within the visible spectrum of electromagnetic radiations. However, all of these magic wavelengths support a red-detuned trap, except 1132.53(6) nm and 1128.42(6) nm for $4S$-$3D_{3/2}$ and $4S - 3D_{5/2}$ transitions, respectively, support far blue-detuned traps and are found to be useful for experimental demonstrations.

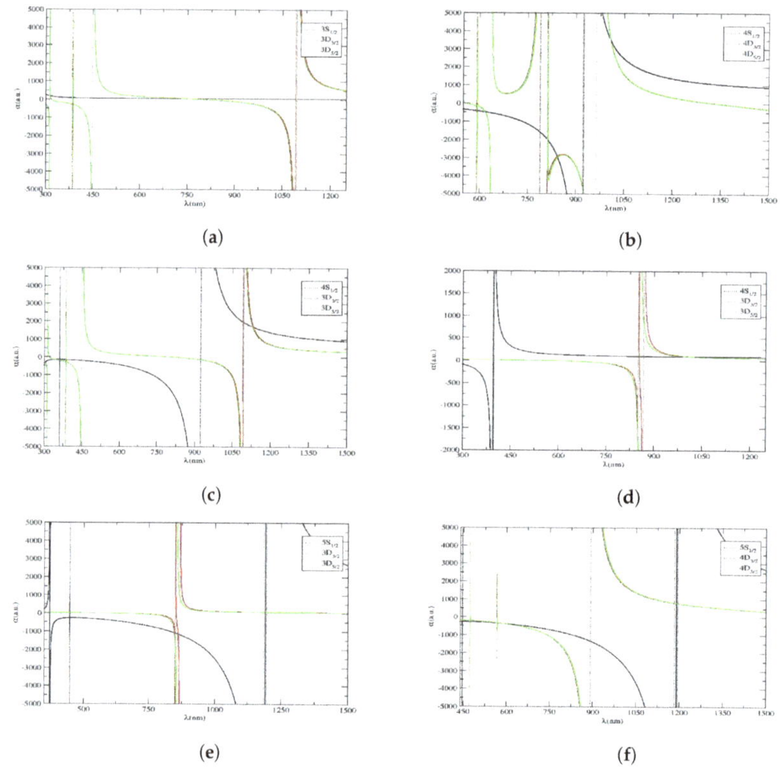

Figure 2. α v/s λ plots for various transitions in Mg^+ and Ca^+ ions. (**a**) α v/s λ plot for $3S_{1/2} - 3D_{3/2,5/2}$ transition in Mg^+ ion. (**b**) α v/s λ plot for $4S_{1/2} - 4D_{3/2,5/2}$ transition in Mg^+ ion. (**c**) α v/s λ plot for $4S_{1/2} - 3D_{3/2,5/2}$ transition in Mg^+ ion. (**d**) α v/s λ plot for $4S_{1/2} - 3D_{3/2,5/2}$ transition in Ca^+ ion. (**e**) α v/s λ plot for $5S_{1/2} - 3D_{3/2,5/2}$ transition in Ca^+ ion. (**f**) α v/s λ plot for $5S_{1/2} - 4D_{3/2,5/2}$ transition in Ca^+ ion.

4.1.2. Ca^+

We considered $4S - 3D_{3/2,5/2}$ and $5S - (4,3)D_{3/2,5/2}$ transitions for locating the magic wavelengths in Ca^+ ion. We tabulated magic wavelengths for these transitions along with the comparison of λ_{magic}s with the only available results for $4S - 3D_{3/2,5/2}$ in Table 2. In addition, we plotted scalar dipole polarizabilities against wavelengths for these transitions in Figure 2d–f, correspondingly. According to Table 2, it is determined that, subsequently, three and two magic wavelengths exist between 393 nm and 1030 nm for $4S - 3D_{3/2,5/2}$ transitions. In both cases, except 1029.0(1.2) nm and 1011.90(90) nm magic wavelengths, which are far-detuned, all other magic wavelengths are close to resonances and are not suitable for laser trapping.

During analysis, six and five magic wavelengths were located for the $5S - (3,4)D_{3/2}$ and $5S - (3,4)D_{5/2}$ transitions, respectively. It was also determined that all the magic wavelengths are approximately same for both $5S - 4D_{3/2}$ and $5S - 4D_{5/2}$ transitions. Moreover, λ_{magic}s around 845 nm, 847 nm and 860 nm share deep trapping potential for blue-detuned traps and, hence, are further recommended for configuring feasible traps. λ_{magic} at 1191.56 nm, identified in the infrared region for both $5S - 4D_{3/2,5/2}$ transitions, is the only magic wavelength that supports the red-detuned trap. In addition, the polarizability for this wavelength is sufficient enough for creating an ion trap at reasonable laser power. To validate our results, we also compared our results with the results provided only for $4S - 3D_{3/2,5/2}$ in Ref. [40], and noticed that the results for these transitions are in good agreement with only less than 1% variation w.r.t. obtained results.

Table 2. Magic wavelengths λ_{magic} (in nm) with their corresponding polarizability $\alpha_n(\omega)$ (in a.u.) for $(4,5)S_{1/2} - 3D_{3/2,5/2}$ and $5S_{1/2} - 4D_{3/2,5/2}$ transitions in Ca^+ ion and their comparison with the available literature.

$4S_{1/2} - 3D_{3/2}$				$4S_{1/2} - 3D_{5/2}$			
Resonance	λ_{res}	λ_{magic}	α_{magic}	**Resonance**	λ_{res}	λ_{magic}	α_{magic}
$4S_{1/2} \rightarrow 4P_{3/2}$	393.48			$4S_{1/2} \rightarrow 4P_{3/2}$	393.47		
		395.795 (3)	5.6 (1.6)			395.795 (3)	5.6 (1.6)
		395.82 (3) [40]	4.90 [40]			395.82 (2) [40]	95.87 [40]
$4S_{1/2} \rightarrow 4P_{1/2}$	396.96			$4S_{1/2} \rightarrow 4P_{1/2}$	396.96		
$3D_{3/2} \rightarrow 4P_{3/2}$	850.04			$3D_{5/2} \rightarrow 4P_{3/2}$	854.44		
		852.42 (2)	95.67 (26)			1011.90 (90)	88.89 (25)
		852.45 (2) [40]	4.20 [40]			1014.10 (3) [40]	89.01 [40]
$3D_{3/2} \rightarrow 4P_{1/2}$	866.45						
		1029.0 (1.2)	88.39 (25)				
		1029.7 (2) [40]	88.55 [40]				
$5S_{1/2} - 3D_{3/2}$				$5S_{1/2} - 3D_{5/2}$			
$4P_{1/2} \rightarrow 5S_{1/2}$	370.71			$4P_{1/2} \rightarrow 5S_{1/2}$	370.71		
		371.76 (1)	6.7 (1.6)			371.76 (1)	6.7 (1.6)
$4P_{3/2} \rightarrow 5S_{1/2}$	373.80			$4P_{3/2} \rightarrow 5S_{1/2}$	373.80		
$5S_{1/2} \rightarrow 6P_{3/2}$	447.33			$5S_{1/2} \rightarrow 6P_{3/2}$	447.33		
		447.385 (5)	3.0 (1.6)			447.385 (5)	2.9 (1.6)
$5S_{1/2} \rightarrow 6P_{1/2}$	448.07			$5S_{1/2} \rightarrow 6P_{1/2}$	448.07		
		448.088 (2)	2.9 (1.6)			448.088 (2)	2.9 (1.6)
		847.69 (2)	−1089.4 (2.8)			845.78 (4)	−1079.8 (2.8)
$3D_{3/2} \rightarrow 4P_{3/2}$	850.04			$3D_{5/2} \rightarrow 4P_{3/2}$	854.44		
		860.22 (7)	−1154.7 (3.0)				
$3D_{3/2} \rightarrow 4P_{1/2}$	866.45						
$5S_{1/2} \rightarrow 5P_{3/2}$	1184.22			$5S_{1/2} \rightarrow 5P_{3/2}$	1184.22		
		1191.59 (1)	58.9 (1.6)		1191.59 (1)		56.8 (1.5)
$5S_{1/2} \rightarrow 5P_{1/2}$	1195.30			$5S_{1/2} \rightarrow 5P_{1/2}$	1195.30		
$5S_{1/2} - 4D_{3/2}$				$5S_{1/2} - 4D_{5/2}$			
$5S_{1/2} \rightarrow 6P_{3/2}$	447.33			$5S_{1/2} \rightarrow 6P_{3/2}$	447.34		
		447.52 (2)	−204.2 (2.4)			447.52 (2)	−205.3 (2.4)
$5S_{1/2} \rightarrow 6P_{1/2}$	448.07			$5S_{1/2} \rightarrow 6P_{1/2}$	448.07		
		448.16 (2)	−204.7 (2.4)			448.16 (1)	−205.8 (2.4)
		471.22 (24)	−279.71 (85)			471.67 (23)	−279.85 (85)
$4D_{3/2} \rightarrow 5F_{5/2}$	471.81			$4D_{5/2} \rightarrow 5F_{5/2,7/2}$	472.23		
		565.51 (2)	−355.55 (91)			566.03 (13)	−356.17 (92)
$4D_{3/2} \rightarrow 6P_{3/2}$	565.53			$4D_{5/2} \rightarrow 6P_{3/2}$	566.15		
		566.67 (17)	−356.93 (92)				
$4D_{3/2} \rightarrow 6P_{1/2}$	566.71						
$4D_{3/2} \rightarrow 4F_{5/2}$	891.45			$4D_{5/2} \rightarrow 4F_{7/2}$	892.98		
$5S_{1/2} \rightarrow 5P_{3/2}$	1184.22			$5S_{1/2} \rightarrow 5P_{3/2}$	1184.22		
		1191.56 (1)	776.1 (8.1)			1191.56 (1)	779.9 (7.9)
$5S_{1/2} \rightarrow 5P_{1/2}$	1195.30			$5S_{1/2} \rightarrow 5P_{1/2}$	1195.302		

4.1.3. Sr^+

Figure 3a–c demonstrate M_J-independent dynamic dipole polarizability versus wavelength plots for $(6,5)S_{1/2} - 4D_{3/2,5/2}$ and $6S_{1/2} - 5D_{3/2,5/2}$ transitions for Sr^+ ion. The results corresponding to these figures are listed in Table 3. Only two magic wavelengths were traced for the $5S - 4D_{3/2}$ transition, whereas only one magic wavelength exists for the $5S - 4D_{5/2}$ transition. According to Table 3, for the $6S - 4D_{3/2}$ transition, three magic wavelengths exist below 480 nm, with a dynamic polarizability of value less than 15 a.u.; however, another three λ_{magic}s, lie between 1000 nm and 1231 nm. The λ_{magic}s at 1002.401 nm and 1087.35 nm support blue-detuned traps with sufficiently high polarizabilities for the experimental trapping of Sr^+ ions. For the $6S - 4D_{5/2}$ transition, five magic wavelengths were located between 420 nm and 1250 nm, out of which the magic wavelengths at 421.47(5) nm, 474.61(1) nm, 477.549(0) nm and 1230.05(6) nm follow red-detuned traps, whereas the only magic wavelength at 1025.19(17) nm, with corresponding $\alpha = -2858.0(8.8)$ a.u., supports a blue-detuned trap, which can be useful for experimental purposes. We recommend this magic wavelength of Sr^+ ion for the $6S$-$4D_{5/2}$ transition. Moreover, it is also observed that all the magic wavelengths for these two transitions lie between the same resonance transitions and are closer to each other. Threfore, it is probable to trap Sr^+ ion for both of these transitions with same magic wavelength.

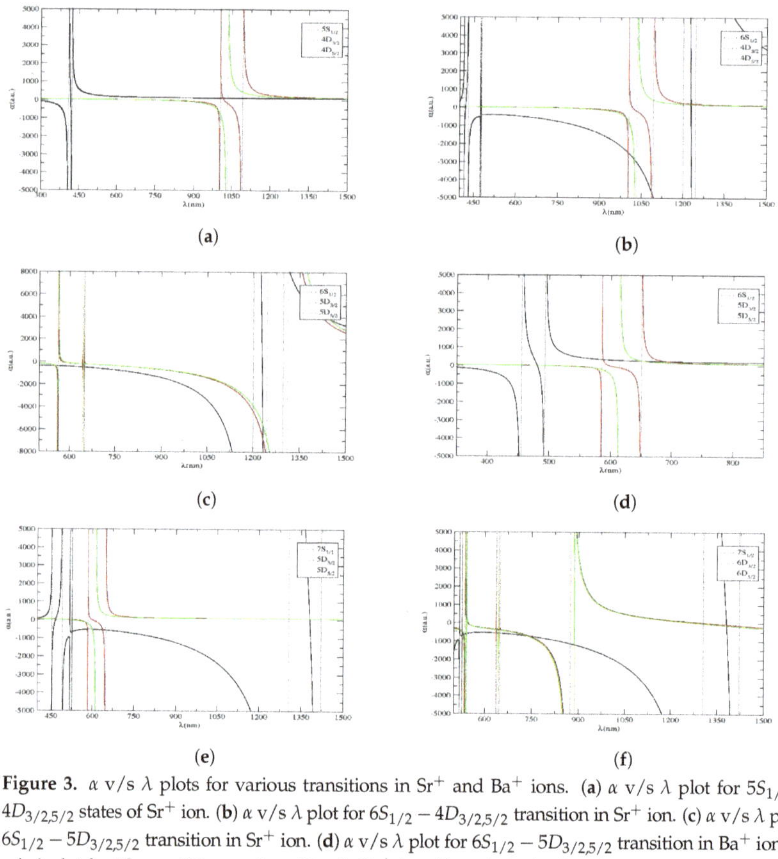

Figure 3. α v/s λ plots for various transitions in Sr^+ and Ba^+ ions. **(a)** α v/s λ plot for $5S_{1/2}$ and $4D_{3/2,5/2}$ states of Sr^+ ion. **(b)** α v/s λ plot for $6S_{1/2} - 4D_{3/2,5/2}$ transition in Sr^+ ion. **(c)** α v/s λ plot for $6S_{1/2} - 5D_{3/2,5/2}$ transition in Sr^+ ion. **(d)** α v/s λ plot for $6S_{1/2} - 5D_{3/2,5/2}$ transition in Ba^+ ion. **(e)** α v/s λ plot for $7S_{1/2} - 5D_{3/2,5/2}$ transition in Ba^+ ion. **(f)** α v/s λ plot for $7S_{1/2} - 6D_{3/2,5/2}$ transition in Ba^+ ion.

Table 3 also shows that there are four magic wavelengths that lie within the wavelength range of 640 nm to 1450 nm for the $6S - 5D_{3/2}$ transition. It is also observed that three out of four magic wavelengths for $6S - 5D_{3/2}$ transition support blue-detuned traps; however, the $\lambda_{magic} = 1233.61(5)$ nm at $\alpha_{magic} = -6756(24)$ a.u. is recommended for experimental purposes as it is far-detuned and a high value of dipole polarizability indicates deep trapping potential. On the other hand, only three magic wavelengths were identified for the $6S - 5D_{5/2}$ transition in Sr^+ ion with two supporting blue-detuned traps. Two out of these λ_{magic}s, i.e., 1233.06(5) nm and 1448.4(1.6) nm, are located at higher wavelength ranges, with deep potentials for their respective favourable blue- and red-detuned traps. Therefore, both of these values are recommended for further experimental studies. Moreover, we compared our magic wavelengths for $5S - 4D_{3/2,5/2}$ transitions with respect to the available literature in the same table. It is seen that our reported values are in excellent approximation with the results obtained by Kaur et al. [40], with a variation by less than 0.05%. Unfortunately, we could not find any data related to other transitions to carry out the comparison with. Hence, it can be concluded from the comparison of available data that our results are promising and can be used for further prospective calculations of atomic structures and atomic properties of this ion.

Table 3. Magic wavelengths λ_{magic} (in nm) with their corresponding polarizability $\alpha_n(\omega)$ (in a.u.) along with their comparison with the available literature for $(5,6)S_{1/2} - 4D_{3/2,5/2}$ and $6S_{1/2} - 5D_{3/2,5/2}$ transitions in Sr^+ ion.

	$5S_{1/2} - 4D_{3/2}$				$5S_{1/2} - 4D_{5/2}$		
Resonance	λ_{res}	λ_{magic}	α_{magic}	**Resonance**	λ_{res}	λ_{magic}	α_{magic}
$5S_{1/2} \rightarrow 5P_{3/2}$	407.89			$5S_{1/2} \rightarrow 5P_{3/2}$	407.89		
		417.00 (5)	15.3 (2.1)			417.00 (4)	15.2 (2.1)
		416.9 (3) [40]	14.47 [40]			416.9 (3) [40]	13.3 [40]
$5S_{1/2} \rightarrow 5P_{1/2}$	421.67			$5S_{1/2} \rightarrow 5P_{1/2}$	421.67		
$4D_{3/2} \rightarrow 5P_{3/2}$	1003.94			$4D_{5/2} \rightarrow 5P_{3/2}$	1003.01		
		1014.68 (26)	108.70 (90)				
		1014.6 (2) [40]	108.35 [40]				
$4D_{3/2} \rightarrow 5P_{1/2}$	1091.79						
	$6S_{1/2} - 4D_{3/2}$				$6S_{1/2} - 4D_{5/2}$		
$5P_{1/2} \rightarrow 6S_{1/2}$	416.27			$5P_{1/2} \rightarrow 6S_{1/2}$	416.30		
		421.47 (5)	15.0 (2.1)			421.47 (5)	14.9 (2.1)
$5P_{3/2} \rightarrow 6S_{1/2}$	430.67			$5P_{3/2} \rightarrow 6S_{1/2}$	430.67		
$6S_{1/2} \rightarrow 7P_{3/2}$	474.37			$6S_{1/2} \rightarrow 7P_{3/2}$	474.37		
		474.61 (1)	11.3 (2.1)			474.61 (1)	10.9 (2.1)
$6S_{1/2} \rightarrow 7P_{1/2}$	477.49			$6S_{1/2} \rightarrow 7P_{1/2}$	477.49		
		477.549 (0)	11.1 (2.1)			477.557 (5)	10.7 (2.1)
		1002.40 (3)	−2470.0 (7.5)			1025.19 (17)	−2858.0 (8.8)
$4D_{3/2} \rightarrow 5P_{3/2}$	1003.94			$4D_{5/2} \rightarrow 5P_{3/2}$	1033.01		
		1087.35 (9)	−4653 (15)				
$4D_{3/2} \rightarrow 5P_{1/2}$	1091.79			$6S_{1/2} \rightarrow 6P_{3/2}$	1201.73		
$6S_{1/2} \rightarrow 6P_{3/2}$	1201.73					1230.05 (6)	170.2 (3.6)
		1230.02 (6)	223.4 (4.2)				
$6S_{1/2} \rightarrow 6P_{1/2}$	1244.84			$6S_{1/2} \rightarrow 6P_{1/2}$	1244.84		
	$6S_{1/2} - 5D_{3/2}$				$6S_{1/2} - 5D_{5/2}$		
$5D_{3/2} \rightarrow 5F_{5/2}$	562.45			$5D_{5/2} \rightarrow 5F_{5/2}$	565.20		
		643.871 (2)	−528.6 (1.5)			647.41 (2)	−534.6 (1.6)
$5D_{3/2} \rightarrow 7P_{3/2}$	643.88			$5D_{5/2} \rightarrow 7P_{3/2}$	647.49		
		649.49 (2)	−538.2 (1.6)				
$5D_{3/2} \rightarrow 7P_{1/2}$	649.65						
$6S_{1/2} \rightarrow 6P_{3/2}$	1201.73			$6S_{1/2} \rightarrow 6P_{3/2}$	1201.73		
		1233.61 (5)	−6756 (29)			1233.06 (5)	−5620 (23)
$6S_{1/2} \rightarrow 6P_{1/2}$	1244.84			$6S_{1/2} \rightarrow 6P_{1/2}$	1244.84		
$5D_{3/2} \rightarrow 4F_{5/2}$	1297.85			$5D_{5/2} \rightarrow 4F_{5/2}$	1312.62		
		1411.9 (1.5)	4382 (13)				
				$5D_{5/2} \rightarrow 4F_{7/2}$	1312.84		
						1448.4 (1.6)	3812 (11)

4.1.4. Ba^+

The results for magic wavelengths for $6S - 5D_{3/2,5/2}$, $7S - 5D_{3/2,5/2}$ and $7S - 6D_{3/2,5/2}$ transitions in Ba^+ ion are tabulated in Table 4. As per Figure 3d and Table 4, a maximum number of magic wavelengths were located between 480 and 700 nm. It is also observed that the magic wavelengths that lie between $6S - 6P_{1/2}$ and $6S - 6P_{3/2}$ resonant transitions support blue-detuned trap; however, the dynamic dipole polarizability corresponding to these magic wavelengths are too small to trap Ba^+ ions at these wavelengths. A total of six magic wavelengths were found for the $7S - 5D_{3/2}$ transition, out of which two lie in the vicinity of 526 nm. The sharp intersection of the polarizability curves of the involved states of transition lie at 583.76(1) nm, 638.75(5) nm and 1380.83(24) nm. Similarly, four magic wavelengths were identified for the $7S - 5D_{5/2}$ transition; however, unlike the $7S - 5D_{3/2}$ transition, no magic wavelength was identified in the vicinity of 600 to 1300 nm. It was also determined that three out of these four λ_{magic}s, support a blue-detuned trap, although the trapping potentials for these traps are not deep enough for further consideration in experimentations.

Table 4. Magic wavelengths λ_{magic} (in nm) with their corresponding polarizability $\alpha_n(\omega)$ (in a.u.) along with their comparison with the available literature for $(6,7)S_{1/2} - 5D_{3/2,5/2}$ and $7S_{1/2} - 6D_{3/2,5/2}$ transitions in Ba^+ ions.

	$6S_{1/2} - 5D_{3/2}$				$6S_{1/2} - 5D_{5/2}$		
Resonance	λ_{res}	λ_{magic}	α_{magic}	**Resonance**	λ_{res}	λ_{magic}	α_{magic}
$6S_{1/2} \to 6P_{3/2}$	455.53			$6S_{1/2} \to 6P_{3/2}$	455.53		
		480.71 (2)	−4.1 (1.6)			480.76 (2)	−8.3 (1.4)
		480.6 (5) [40]	−2.89 [40]				
$6S_{1/2} \to 6P_{1/2}$	493.55			$6S_{1/2} \to 6P_{1/2}$	493.55		
$5D_{3/2} \to 6P_{3/2}$	585.53			$5D_{3/2} \to 6P_{3/2}$	614.34		
		588.32 (1)	330.15 (60)			653.17 ()35	247.90 (57)
		588.4 (3) [40]	329.33 [40]			695.7 (3) [40]	219.4 [40]
$5D_{3/2} \to 6P_{1/2}$	649.87						
		693.46 (48)	221.91 (56)				
		655.50 (3) [40]	244.89 [40]				
	$7S_{1/2} - 5D_{3/2}$				$7S_{1/2} - 5D_{5/2}$		
$6P_{1/2} \to 7S_{1/2}$	452.62			$6P_{1/2} \to 7S_{3/2}$	452.62		
		466.95 (13)	0.5 (1.6)			466.88 (9)	−2.6 (1.5)
$6P_{3/2} \to 7S_{1/2}$	490.13			$6P_{3/2} \to 7S_{1/2}$	490.13		
$7S_{1/2} \to 8P_{3/2}$	518.49			$7S_{1/2} \to 8P_{3/2}$	518.49		
		518.79 (2)	−22.9 (1.5)			518.79 (2)	−31.9 (1.5)
$7S_{1/2} \to 8P_{1/2}$	526.75			$7S_{1/2} \to 8P_{1/2}$	526.75		
		526.779 (6)	−28.7 (1.5)			601.37 (6)	−548.8 (1.6)
		583.76 (1)	−548.6 (2.4)				
$5D_{3/2} \to 6P_{3/2}$	585.53			$5D_{5/2} \to 6P_{3/2}$	614.34		
		638.75 (5)	−573.9 (2.4)				
$5D_{3/2} \to 6P_{1/2}$	649.87						
$7S_{1/2} \to 7P_{3/2}$	1306.14			$7S_{1/2} \to 7P_{3/2}$	1306.14		
		1380.83 (24)	59.8 (1.4)			1380.83 (24)	59.2 ()1.3
$7S_{1/2} \to 7P_{1/2}$	1421.54			$7S_{1/2} \to 7P_{1/2}$	1421.54		
	$7S_{1/2} - 6D_{3/2}$				$7S_{1/2} - 6D_{5/2}$		
$7S_{1/2} \to 8P_{3/2}$	518.49			$7S_{1/2} \to 8P_{3/2}$	518.49		
		519.03 (11)	−376 (34)			519.07 (10)	−405 (31)
$7S_{1/2} \to 8P_{1/2}$	526.750			$7S_{1/2} \to 8P_{1/2}$	526.75		
		526.84 (12)	−483 (43)			526.84 (6)	−480 (43)
		531.0 (1.0)	−664.2 (4.0)			532.8 (2.4)	−655.2 (3.8)
$6D_{3/2} \to 6F_{5/2}$	536.28			$6D_{5/2} \to 6F_{7/2}$	539.31		
						542.17 (7)	−613.1 (3.2)
				$6D_{5/2} \to 6F_{5/2}$	542.26		
						645.33 (10)	−580.8 (2.5)
$6D_{3/2} \to 8P_{3/2}$	637.25			$6D_{5/2} \to 8P_{3/2}$	645.70		
						735.65 (92)	−728.3 (3.2)
$6D_{3/2} \to 8P_{1/2}$	649.77			$6D_{5/2} \to 5F_{7/2}$	871.32		
		744.0 (2.2)	−746.4 (3.2)			889.20 (20)	−1211.6 (5.5)
$6D_{3/2} \to 5F_{5/2}$	874.02			$6D_{5/2} \to 5F_{5/2}$	889.99		
$7S_{1/2} \to 7P_{3/2}$	1306.14			$7S_{1/2} \to 7P_{3/2}$	1306.14		
		1381.25 (35)	−78 (77)			1381.39 (33)	−125 (77)
$7S_{1/2} \to 7P_{1/2}$	1421.54			$7S_{1/2} \to 7P_{1/2}$	1421.54		

Table 4 also compiles the magic wavelengths for $7S - 5D_{3/2,5/2}$ transitions and shows that there exists six and four magic wavelengths for $7S - 5D_{3/2}$ and $7S - 5D_{5/2}$ transition, respectively. It is also seen that the magic wavelengths between $6P_{3/2} - 7S$ and $7S - 8P_{3/2}$ as well as the $5D_{3/2} - 6P_{1/2}$ and $7S - 7P_{3/2}$ transitions seem to be missing, as shown in Figure 3e. It is also observed that the magic wavelength at 466.95(13) nm and 1380.83(24) nm are slightly red-shifted; nevertheless, the λ_{magic} at 638.75(5) nm in the visible region supports a blue-detuned trap, and can have sufficient trap depth at reasonable laser power.

Similarly, the magic wavelengths and their corresponding dynamic dipole polarizability along with their comparison with available literature is also provided in the same table for $7S - 6D_{3/2,5/2}$ transitions. The same was demonstrated graphically in Figure 3f, which includes a total of thirteen magic wavelengths in all for the considered transitions. It is also examined that no magic wavelength exists between $6D_{3/2} - 6F_{5/2}$ and $6D_{3/2} - 8P_{1/2}$ resonances. Unlike the $7S - 6D_{3/2}$ transition, around eight magic wavelengths were located between $7S - 8P_{3/2}$ and $7S - 7P_{1/2}$ resonances, and all of them support blue-detuned traps. Moreover, magic wavelengths at 532.8(2.4) nm, 735.65(92) nm and 1381.39(33) nm are expected to be more promising for experiments due to sufficient trap depths for reasonable-power lasers. However, on the comparison of our results for $6S - 5D_{3/2,5/2}$ transitions for Ba^+ ion, we observed that all the magic wavelengths agree well with the results obtained by Kaur et al., in Ref. [40], except the last magic wavelengths that were identified at 693 nm and 653 nm for $6S - 5D_{3/2}$ and $6S - 5D_{5/2}$ transitions.

4.2. Tune-Out Wavelengths

We illustrated tune-out wavelengths for different states of the considered transitions in the alkaline-earth ions along with their comparison with the already available literature in Table 5. To locate these M_J-independent tune-out wavelengths, we evaluated the scalar dipole dynamic polarizabilities of these states for considered alkaline-earth ions and identified those values of λ for which polarizability vanished. It is also highlighted that, in Mg^+ ion, all the tune-out wavelengths identified for $3S_{1/2}$ and $4S_{1/2}$ states lie in the UV region, whereas, for $(3,4)D_{3/2,5/2}$ states, a few tune-out wavelengths are located in visible range. Moreover, the largest λ_T is identified for the $4D_{3/2}$ state at 1331.53(53) nm. Furthermore, only one tune-out wavelength, i.e., $\lambda_T = 280.1136(2)$ nm for $3S_{1/2}$, could be compared with the result presented by Kaur et al. in Ref. [55] and it is seen that our result is in good accord with this value.

Similarly, we pointed out tune-out wavelengths for $nS_{1/2}$ and $(n-1)D_{3/2}$, $n = (4,5), (5,6)$ and $(6,7)$ states for Ca^+, Sr^+ and Ba^+ ions, by identifying λs at which their corresponding αs tend to zero. Hence, it was perceived that, out of 25 tune-out wavelengths for all states of Ca^+ ion, only seven of them lie within visible spectrum and, on comparison of different tune-out wavelengths for the $4S_{1/2}$ and $3D_{3/2}$ states of Ca^+ ion, it was determined that all of these results are supported by the results obtained in Refs. [40,55]. However, one of the tune-out wavelength located at 493.13(87) nm for the $3D_{5/2}$ state of Ca^+ ion seems have 2% variation from the wavelength obtained by Kaur et al. in Ref. [40]. This results from the fact that the present study incorporates more precise E1 matrix elements for the high-lying transitions as well their excitation energies from the portal for High-Precision Atomic Data and Computation [54].

For Sr^+ ion, the maximum number of tune-out wavelengths was identified out of all the considered alkaline-earth ions. It is also realized that most of these λs lie within the visible spectrum of electromagnetic radiation, and are mostly comprised of all the λ_T values corresponding to $5S_{1/2}$, $5D_{3/2}$ and $5D_{5/2}$ states. Additionally, during the comparison of these values with the results published in Refs. [40,55], it was determined that the tune-out wavelength at 417.04(6) nm for $5S_{1/2}$ as well as $\lambda_T = 1018.91(38)$ nm for the $4D_{3/2}$ state agree well with the available results; howsoever, the tune-out wavelengths at 606.47(53) nm and 593.0(2.0) nm for $4D_{3/2}$ and $4D_{5/2}$ states, respectively, show a discrepancy of less than 2%, which lies within the error bar of 5%.

In the case of Ba^+ ion, we located 23 tune-out wavelengths, which, in all, comprise of 10, 9 and 4 wavelengths in visible, UV and infrared regions, respectively. It is also highlighted that all the tune-out wavelengths that exist in visible region lie within the range 480 nm to 550 nm. We also compared our tune-out wavelengths for $6S_{1/2}$ and $5D_{3/2,5/2}$ states against available theoretical data in Refs. [38,40,55] and it was found that all the λ_Ts except 468.61 nm and 459.57 nm, respectively, for $5D_{3/2}$ and $5D_{5/2}$ states show disparities of less than 1%, which lies within the considerable error limit.

Table 5. Tune-out wavelengths λ_T (in nm) of various states of Mg^+, Ca^+, Sr^+ and Ba^+ ions and their comparison with available literature.

Mg⁺			Ca⁺			Sr⁺			Ba⁺		
State	λ_T	Others	State	λ_T	Others	State	λ_T	Others	State	λ_T	Others
$3S_{1/2}$	102.6058 (6) 102.696 (8) 124.071 (7) 280.1136 (2)	280.110 (9) [55]	$4S_{1/2}$	165.04 (3) 165.26 (4) 395.796 (3)	395.80 (2) [55] 395.796 [40]	$5S_{1/2}$	417.04 (5)	417.04 (6) [55] 417.025 [40]	$6S_{1/2}$	200.035 (1) 202.51 (9) 480.66 (2)	480.63 (24) [55] 480.66 (18) [38] 480.596 [40]
$3D_{3/2}$	317.082 (59) 384.9615 (4) 385.3401 (9) 812.03 (97) 1092.436 (9)		$3D_{3/2}$	212.930 (5) 213.25 (2) 494.37 (63) 852.75 (3)	492.752 [40] 852.776 [40]	$4D_{3/2}$	185.498 (2) 192.95 (95) 242.622 (7) 606.47 (53) 1018.91 (38)	598.633 [40] 1018.873 [40]	$5D_{3/2}$	224.675 (6) 468.61 (61) 597.93 (15)	472.461 [40] 597.983 [40]
$3D_{5/2}$	317.003 (6) 385.143 (9) 810.69 (31)		$3D_{5/2}$	170.63 (63) 213.16 (3) 493.13 (87)	482.642 [40]	$4D_{5/2}$	193.78 (78) 242.56 (1) 593.0 (2.0)	585.677 [40]	$5D_{5/2}$	193.32 (16) 198.43 (57) 225.63 (1) 234.773 (3) 459.57 (57)	509.687 [40]
$4S_{1/2}$	279.1332 (4) 293.2134 (2) 361.524 (2) 361.691 (1)		$5S_{1/2}$	287.179 (7) 299.64 (64) 309.13 (5) 309.303 (9) 371.754 (9) 447.386 (5) 448.088 (2) 1191.59 (1)		$6S_{1/2}$	334.71 (71) 362.58 (2) 378.373 (3) 421.40 (5) 474.62 (1) 477.559 (5) 1230.15 (6)		$7S_{1/2}$	347.63 (5) 363.36 (64) 394.63 (6) 466.94 (13) 518.78 (2) 526.778 (6) 1381.01 (23)	
$4D_{3/2}$	554.11 (12) 592.70 (3) 1331.53 (53)		$4D_{3/2}$	317.850 (5) 334.65 (1) 371.2822 (5) 375.78 (3) 471.96 (6)		$5D_{3/2}$	407.4377 (3) 434.69 (23) 481.52 (2) 573.80 (56) 649.82 (2)		$6D_{3/2}$	544.38 (62) 1316.93 (93)	
$4D_{5/2}$	507.25 (1) 553.95 (95) 592.55 (3) 811.9701 (4) 1329.45 (55)		$4D_{5/2}$	376.05 (3) 472.38 (6) 566.15 (10)		$5D_{5/2}$	436.25 (22) 455.827 (4) 577.18 (57) 647.60 (2) 1312.6247 (1)		$6D_{5/2}$	547.53 (53) 889.04 (23) 1287.17 (83)	

5. Conclusions

We identified a number of reliable magnetic-sublevel-independent tune-out wavelengths of many $S_{1/2}$ and $D_{3/2,5/2}$ states, and magic wavelengths of different combinations of $S_{1/2} - D_{3/2,5/2}$ transitions in the alkaline-earth ions from Mg^+ through Ba^+. If they can be measured precisely, accurate values of many electric dipole matrix elements can be inferred by combining the experimental values of these quantities with our theoretical results. Most of the magic wavelengths found from this study show that they can be detected using the red- and blue-detuned traps. In fact, it is possible to perform many high-precision measurements by trapping the atoms at the reported tune-out and magic wavelengths of the considered transitions in the future, which can be applied to different metrological studies.

Author Contributions: B.A. conceived the idea of magnetic-sublevel-independent magic and tune-out wavelengths for alkaline-earth ions which motivated J. to carry out this study. J. identified magic and tune-out wavelengths for the considered transitions in alkaline-earth ions and drafted the manuscript. H.K. verified the results and contributed in the editing of the manuscript. B.K.S. provided critical feedback and helped shape the research, analysis and manuscript. He also aided in interpretation and discussion of the results and commented on the manuscript. With the help of B.A., H.K. and B.K.S., J. modified this manuscript to the final version. All authors have read and agreed to the published version of the manuscript.

Funding: The work of B.A. is supported by SERB-TARE (TAR/2020/000189), New Delhi, India.

Institutional Review Board Statement: Not Applicable.

Informed Consent Statement: Not Applicable.

Data Availability Statement: The data underlying this article are available in the article.

Acknowledgments: Special thanks to the support of SERB-TARE (TAR/2020/000189), New Delhi, India.The employed relativistic many-body method was developed in the group of Professor M. S. Safronova of the University of Delaware, USA.

Conflicts of Interest: The authors declare no conflict of interest.

References

1. Hall, J.L.; Zhu, M.; Buch, P. Prospects for using laser-prepared atomic fountains for optical frequency standards applications. *JOSA B* **1989**, *6*, 2194–2205. [CrossRef]
2. Kostelecký, V.A.; Vargas, A.J. Lorentz and C P T tests with clock-comparison experiments. *Phys. Rev. D* **2018**, *98*, 036003. [CrossRef]
3. Wood, C.; Bennett, S.; Cho, D.; Masterson, B.; Roberts, J.; Tanner, C.; Wieman, C. Measurement of parity nonconservation and an anapole moment in cesium. *Science* **1997**, *275*, 1759–1763. [CrossRef]
4. Tiecke, T.; Thompson, J.D.; de Leon, N.P.; Liu, L.; Vuletić, V.; Lukin, M.D. Nanophotonic quantum phase switch with a single atom. *Nature* **2014**, *508*, 241–244. [CrossRef]
5. Xiaxing, X.; Mouqi, H.; Youyuan, Z.; Zhiming, Z. The parity non-conservation E1 matrix of barium-a semi-empirical calculation. *J. Phys. B Mol. Opt. Phys.* **1990**, *23*, 4239. [CrossRef]
6. Berengut, J.C.; Dzuba, V.A.; Flambaum, V.V. Isotope-shift calculations for atoms with one valence electron. *Phys. Rev. A* **2003**, *68*, 022502. [CrossRef]
7. Fortier, T.M.; Ashby, N.; Bergquist, J.C.; Delaney, M.J.; Diddams, S.A.; Heavner, T.P.; Hollberg, L.; Itano, W.M.; Jefferts, S.R.; Kim, K.; et al. Precision Atomic Spectroscopy for Improved Limits on Variation of the Fine Structure Constant and Local Position Invariance. *Phys. Rev. Lett.* **2007**, *98*, 070801. [CrossRef]
8. Roos, C.F.; Riebe, M.; Haffner, H.; Hansel, W.; Benhelm, J.; Lancaster, G.P.; Becher, C.; Schmidt-Kaler, F.; Blatt, R. Control and measurement of three-qubit entangled states. *Science* **2004**, *304*, 1478–1480. [CrossRef]
9. Kajita, M.; Li, Y.; Matsubara, K.; Hayasaka, K.; Hosokawa, M. Prospect of optical frequency standard based on a $^{43}Ca^+$ ion. *Phys. Rev. A* **2005**, *72*, 043404. [CrossRef]
10. Enderlein, M.; Huber, T.; Schneider, C.; Schaetz, T. Single ions trapped in a one-dimensional optical lattice. *Phys. Rev. Lett.* **2012**, *109*, 233004. [CrossRef]
11. Chou, C.w.; Hume, D.; Koelemeij, J.; Wineland, D.J.; Rosenband, T. Frequency comparison of two high-accuracy Al+ optical clocks. *Phys. Rev. Lett.* **2010**, *104*, 070802. [CrossRef] [PubMed]
12. Champenois, C.; Marciante, M.; Pedregosa-Gutierrez, J.; Houssin, M.; Knoop, M.; Kajita, M. Ion ring in a linear multipole trap for optical frequency metrology. *Phys. Rev. A* **2010**, *81*, 043410. [CrossRef]

13. Katori, H.; Ido, T.; Kuwata-Gonokami, M. Optimal design of dipole potentials for efficient loading of Sr atoms. *J. Phys. Soc. Jpn.* **1999**, *68*, 2479–2482. [CrossRef]
14. Liu, P.L.; Huang, Y.; Bian, W.; Shao, H.; Guan, H.; Tang, Y.B.; Li, C.B.; Mitroy, J.; Gao, K.L. Measurement of Magic Wavelengths for the $^{40}Ca^+$ Clock Transition. *Phys. Rev. Lett.* **2015**, *114*, 223001. [CrossRef]
15. Huang, Y.; Guan, H.; Li, C.; Zhang, H.; Zhang, B.; Wang, M.; Tang, L.; Shi, T.; Gao, K. Measurement of infrared magic wavelength for an all-optical trapping $^{40}Ca^+$ ion clock. *arXiv* **2022**, arXiv:2202.07828.
16. Kaur, J.; Singh, S.; Arora, B.; Sahoo, B. Magic wavelengths in the alkaline-earth-metal ions. *Phys. Rev. A* **2015**, *92*. [CrossRef]
17. Schneider, C.; Enderlein, M.; Huber, T.; Schätz, T. Optical trapping of an ion. *Nat. Photonics* **2010**, *4*, 772–775. [CrossRef]
18. Sackett, C.; Kielpinski, D.; King, B.; Langer, C.; Meyer, V.; Myatt, C.; Rowe, M.; Turchette, Q.; Itano, W.; Wineland, D.; et al. Experimental entanglement of four particles. *Nature* **2000**, *404*, 256–259. [CrossRef]
19. Tang, Y.B.; Qiao, H.X.; Shi, T.y.; Mitroy, J. Dynamic polarizabilities for the low lying states of Ca+. *Phys. Rev. A* **2013**, *87*, 042517. [CrossRef]
20. Ruffoni, M.; Den Hartog, E.; Lawler, J.; Brewer, N.; Lind, K.; Nave, G.; Pickering, J. Fe I oscillator strengths for the Gaia-ESO survey. *Mon. Not. R. Astron. Soc.* **2014**, *441*, 3127–3136. [CrossRef]
21. Wittkowski, M. Fundamental stellar parameters Technology roadmap for future interferometric facilities. In *Bulletin de la Sociètè Royale des Sciences de Liège, Proceedings of the European Interferometry Initiative Workshop Organized in the Context of the 2005 Joint European and National Astronomy Meeting "Distant Worlds", Liège, Belgium, 6–8 July 2005;* Surdej, J., Caro, D., Detal, A. , Eds.; Liège University, Institute of Astrophysics: Liège, Belgium, 2005; Volume 74, pp. 165–181.
22. LeBlanc, L.; Thywissen, J. Species-specific optical lattices. *Phys. Rev. A* **2007**, *75*, 053612. [CrossRef]
23. Catani, J.; Barontini, G.; Lamporesi, G.; Rabatti, F.; Thalhammer, G.; Minardi, F.; Stringari, S.; Inguscio, M. Entropy Exchange in a Mixture of Ultracold Atoms. *Phys. Rev. Lett.* **2009**, *103*, 140401. [CrossRef] [PubMed]
24. Wang, Y.; Zhang, X.; Corcovilos, T.A.; Kumar, A.; Weiss, D.S. Coherent Addressing of Individual Neutral Atoms in a 3D Optical Lattice. *Phys. Rev. Lett.* **2015**, *115*, 043003. [CrossRef] [PubMed]
25. Kotochigova, S.; Tiesinga, E. Controlling polar molecules in optical lattices. *Phys. Rev. A* **2006**, *73*, 041405. [CrossRef]
26. Rubio-Abadal, A.; Choi, J.y.; Zeiher, J.; Hollerith, S.; Rui, J.; Bloch, I.; Gross, C. Many-Body Delocalization in the Presence of a Quantum Bath. *Phys. Rev. X* **2019**, *9*, 041014. [CrossRef]
27. Holmgren, W.F.; Trubko, R.; Hromada, I.; Cronin, A.D. Measurement of a Wavelength of Light for Which the Energy Shift for an Atom Vanishes. *Phys. Rev. Lett.* **2012**, *109*, 243004. [CrossRef]
28. Herold, C.D.; Vaidya, V.D.; Li, X.; Rolston, S.L.; Porto, J.V.; Safronova, M.S. Precision Measurement of Transition Matrix Elements via Light Shift Cancellation. *Phys. Rev. Lett.* **2012**, *109*, 243003. [CrossRef]
29. Petrov, A.; Makrides, C.; Kotochigova, S. External field control of spin-dependent rotational decoherence of ultracold polar molecules. *Mol. Phys.* **2013**, *111*, 1731–1737. [CrossRef]
30. Henson, B.M.; Khakimov, R.I.; Dall, R.G.; Baldwin, K.G.H.; Tang, L.Y.; Truscott, A.G. Precision Measurement for Metastable Helium Atoms of the 413 nm Tune-Out Wavelength at Which the Atomic Polarizability Vanishes. *Phys. Rev. Lett.* **2015**, *115*, 043004. [CrossRef]
31. Kao, W.; Tang, Y.; Burdick, N.Q.; Lev, B.L. Anisotropic dependence of tune-out wavelength near Dy 741-nm transition. *Opt. Express* **2017**, *25*, 3411–3419. [CrossRef]
32. Heinz, A.; Park, A.J.; Šantić, N.; Trautmann, J.; Porsev, S.G.; Safronova, M.S.; Bloch, I.; Blatt, S. State-Dependent Optical Lattices for the Strontium Optical Qubit. *Phys. Rev. Lett.* **2020**, *124*, 203201. [CrossRef] [PubMed]
33. Arora, B.; Safronova, M.; Clark, C.W. Tune-out wavelengths of alkali-metal atoms and their applications. *Phys. Rev. A* **2011**, *84*, 043401. [CrossRef]
34. Bause, R.; Li, M.; Schindewolf, A.; Chen, X.Y.; Duda, M.; Kotochigova, S.; Bloch, I.; Luo, X.Y. Tune-Out and Magic Wavelengths for Ground-State $^{23}Na^{40}K$ Molecules. *Phys. Rev. Lett.* **2020**, *125*, 023201. [CrossRef] [PubMed]
35. Jiang, J.; Jiang, L.; Wang, X.; Zhang, D.H.; Xie, L.Y.; Dong, C.Z. Magic wavelengths of the Ca^+ ion for circularly polarized light. *Phys. Rev. A* **2017**, *96*, 042503. [CrossRef]
36. Jiang, J.; Jiang, L.; Wang, X.; Shaw, P.; Zhang, D.H.; Xie, L.Y.; Dong, C.Z. Magic wavelengths of Ca^+ ion for linearly and circularly polarized light. *J. Physics Conf. Ser.* **2017**, *875*, 122003. [CrossRef]
37. Chanu, S.R.; Koh, V.P.W.; Arnold, K.J.; Kaewuam, R.; Tan, T.R.; Zhang, Z.; Safronova, M.S.; Barrett, M.D. Magic wavelength of the Ba+138 6s S1/22–5d D5/22 clock transition. *Phys. Rev. A* **2020**, *101*, 042507. [CrossRef]
38. Jiang, J.; Ma, Y.; Wang, X.; Dong, C.Z.; Wu, Z. Tune-out and magic wavelengths of Ba^+ ions. *Phys. Rev. A* **2021**, *103*, 032803. [CrossRef]
39. Singh, S.; Sahoo, B.; Arora, B. Magnetic-sublevel-independent magic wavelengths: Application to Rb and Cs atoms. *Phys. Rev. A* **2016**, *93*, 063422. [CrossRef]
40. Kaur, J.; Singh, S.; Arora, B.; Sahoo, B. Annexing magic and tune-out wavelengths to the clock transitions of the alkaline-earth-metal ions. *Phys. Rev. A* **2017**, *95*, 042501. [CrossRef]
41. Sherman, J.A.; Koerber, T.W.; Markhotok, A.; Nagourney, W.; Fortson, E.N. Precision Measurement of Light Shifts in a Single Trapped Ba^+ Ion. *Phys. Rev. Lett.* **2005**, *94*, 243001. [CrossRef]
42. Sahoo, B.K.; Wansbeek, L.W.; Jungmann, K.; Timmermans, R.G.E. Light shifts and electric dipole matrix elements in Ba^+ and Ra^+. *Phys. Rev. A* **2009**, *79*, 052512. [CrossRef]

43. Beloy, K. *Theory of the Ac Stark Effect on the Atomic Hyperfine Structure and Applications to Microwave Atomic Clocks*; University of Nevada: Reno, NV, USA, 2009.
44. Blundell, S.; Johnson, W.; Sapirstein, J. Relativistic all-order calculations of energies and matrix elements in cesium. *Phys. Rev. A* **1991**, *43*, 3407. [CrossRef] [PubMed]
45. Safronova, M.; Johnson, W. All-order methods for relativistic atomic structure calculations. *Adv. At. Mol. Opt. Phys.* **2008**, *55*, 191–233.
46. Sahoo, B.; Nandy, D.; Das, B.; Sakemi, Y. Correlation trends in the hyperfine structures of Fr 210, 212. *Phys. Rev. A* **2015**, *91*, 042507. [CrossRef]
47. Sahoo, B.; Das, B. Theoretical studies of the long lifetimes of the 6 d D 3/2, 5/2 2 states in Fr: Implications for parity-nonconservation measurements. *Phys. Rev. A* **2015**, *92*, 052511. [CrossRef]
48. Safronova, M.; Johnson, W.; Derevianko, A. Relativistic many-body calculations of energy levels, hyperfine constants, electric-dipole matrix elements, and static polarizabilities for alkali-metal atoms. *Phys. Rev. A* **1999**, *60*, 4476. [CrossRef]
49. Iskrenova-Tchoukova, E.; Safronova, M.S.; Safronova, U. High-precision study of Cs polarizabilities. *J. Comput. Methods Sci. Eng.* **2007**, *7*, 521 540. [CrossRef]
50. Safronova, U.; Johnson, W.; Safronova, M. Excitation energies, polarizabilities, multipole transition rates, and lifetimes of ions along the francium isoelectronic sequence. *Phys. Rev. A* **2007**, *76*, 042504. [CrossRef]
51. Kaur, M.; Dar, D.F.; Sahoo, B.; Arora, B. Radiative transition properties of singly charged magnesium, calcium, strontium and barium ions. *At. Data Nucl. Data Tables* **2020**, *137*, 101381. [CrossRef]
52. Arora, B.; Safronova, M.; Clark, C. Magic wavelengths for the np-ns transitions in alkali-metal atoms. *Phys. Rev. A* **2007**, *76*, 052509. [CrossRef]
53. Ralchenko, Y. NIST atomic spectra database. *Mem. Della Soc. Astron. Ital. Suppl.* **2005**, *8*, 96.
54. Barakhshan, P.; Marrs, A.; Bhosale, A.; Arora, B.; Eigenmann, R.; Safronova, M.S. *Portal for High-Precision Atomic Data and Computation*; (Version 2.0); University of Delaware: Newark, DE, USA. Available online: https://www.udel.edu/atom (accessed on 27 March 2022).
55. Kaur, M.; Singh, S.; Sahoo, B.; Arora, B. Tune-out and magic wavelengths, and electric quadrupole transition properties of the singly charged alkaline-earth metal ions. *At. Data Nucl. Data Tables* **2021**, *140*, 101422. [CrossRef]

atoms

MDPI

Article

Semiempirical Calculations on Low-Energy Electron Scattering by Zn and Cd Atoms

Felipe Arretche [1,*], Wagner Tenfen [2] and Bijaya K. Sahoo [3]

[1] Physics Department, Universidade Federal de Santa Catarina, Florianópolis 88040-900, SC, Brazil
[2] Physics Department, Universidade Federal de Pelotas, Pelotas 96010-900, RS, Brazil; wtenfen@ufpel.edu.br
[3] Atomic, Molecular and Optical Physics Division, Physical Research Laboratory, Navrangpura, Ahmedabad 380058, India; bijaya@prl.res.in
* Correspondence: f.arretche@ufsc.br

Abstract: Since total cross section measurements for electron scattering by Zn and Cd performed in the 1970s, the existence of p-wave shape resonances below 1 eV are well established in the literature. It was suggested that a second d-wave shape resonance could exist in both systems at an energy slightly higher than the one recorded for the p-wave but still below the inelastic threshold. We report elastic scattering calculations for electron collisions with Zn and Cd atoms below 4 eV using a semiempirical approach, as well the scattering length for both targets. Our results show that, indeed, the d-wave shape resonance is found in Zn but absent in Cd. In fact, our cross sections and the few other ones available for this energy range are in discrepancy with the available experimental total cross sections for Cd.

Keywords: elastic scattering; electron scattering; semiempirical potentials; shape resonances

PACS: 34.80.Bm; 34.80.-i

Citation: Arretche, F.; Tenfen, W.; Sahoo, B.K. Semiempirical Calculations on Low-Energy Electron Scattering by Zn and Cd Atoms. *Atoms* **2022**, *10*, 69. https://doi.org/10.3390/atoms10030069

Academic Editors: Rajesh Srivastava and Dmitry V. Fursa

Received: 30 May 2022
Accepted: 22 June 2022
Published: 29 June 2022

1. Introduction

The study of electron–Zn and electron–Cd collisions may provide valuable information for the modeling of metal vapor plasmas [1]. In the same spirit, since electron–atom collisions constitute the most basic chemical reaction that an atom can go through, the characterization of electronic collisions with metal vapor atoms constitutes a problem with its own relevance within atomic physics.

In spite of several decades of theoretical and experimental investigations on electron–Zn/Cd scattering, accurate determination of the elastic and inelastic cross sections is far from being considered as a closed problem, mainly in the low-energy domain. Here, we focus on energies below ≈4.0 eV where the scattering is purely elastic for both the targets.

Surprisingly, despite almost 100 years since the first measurements of electron–Zn/Cd cross sections [2,3], we find a limited number of published works on the determination of total cross sections for energies below 4 eV. In an article from 1976, Burrow et al. [4] reported measurements of low-energy electron scattering by Zn and Cd using an electron transmission method. Due to the limitations intrinsic to the measurement process, only transmitted currents in arbitrary units as a function of the incident energy were presented. The experimental data showed the existence of shape resonances which were identified with the $(ns^2np)^2P$ ground state configurations of the negative ions with energies (widths) 0.49 (0.45) and 0.33 (0.33) eV for Zn and Cd, respectively. In 1991, Marinković et al. [5] presented measurements for the relative differential cross sections (DCSs) for elastic and electronic excitation cross sections for Cd from 3.4 to 85 eV and, in 2002, Kontros et al. [6] presented total cross section data for electron scattering by Cd with resonant structures being found at 0.33 and 3.74 eV. In 2003, Sullivan et al. [7] published an extensive investigation focused on the study of the formation of transient ionic states in electron–Zn/Cd collisions, however, total cross sections for energies lower than 4 eV were not effectively presented.

From the theoretical side, the scenario is not much different. In 1992, Yuan and Zhang [8] reported elastic cross sections for Zn and Cd calculated based on the model correlation potential of Perdew and Zunger (PZ) [9] (as recommended by Padial and Norcross [10]). From now on, they shall be denoted as model potential (MP) calculations. In 2005, Zatsarinny and Bartschat [11] presented benchmark *ab initio* calculations for low-energy electron–Zn scattering performed with the R-matrix method. In addition to these two references, we find the *ab initio* calculations of Berrington et al. [12] for electron–Cd. This work was performed with four different theoretical methodologies and can be considered as the benchmark for a Cd analog to the one of Zatsarinny and Bartschat [11] for Zn. Due to the energy range of interest in this study, we direct special attention to the DCSs at 3.4 eV calculated with the convergent close-coupling (CCC) and the relativistic convergent close-coupling (RCCC) because they allow a direct contrast between the relativistic- and non-relativistic-based calculations. Finally, we have the recent article of McEachran et al. [13] for electron–Zn where cross sections from 0.01 to 5000 eV were recommended for transport simulations.

In addition to the unequivocal presence of the p-wave shape resonances, Burrow et al. [4] conjectured about the possible existence of resonances of the same nature in the d-wave in both atoms. In their own words: "There is faint evidence for a very broad feature in each cross section between the p-wave shape resonance and the first excited states of the neutral. It is tempting but entirely speculative to suggest that this is associated with a d-wave shape resonance". This point is discussed in more detail in the Buckman and Clark review article [14] where the d-wave resonances are guessed at 2.5 and 2.0 eV for Zn and Cd, respectively.

As we will see in Section 3, in the case of Zn there is good agreement between the MP [8] and the *ab initio* R-matrix calculations [11] for energies above $\approx$1 eV. Nevertheless, some discrepancy exists in the description of the position and width of the p-wave resonance. Additionally, no d-wave resonance is found in either calculation. For Cd, there is great divergence among the MP cross sections [8] and the experimental data of Kontros et al. [6]. Given our recent experience investigating shape resonances in positron–Zn/Cd scattering at low energies [15], we understand that further investigations on the possible existence of d-wave shape resonances would be welcome as well as a second theoretical calculation for Cd in order to contrast with the previous theoretical and experimental results.

Given these considerations and the limited set of information available to study the subject, at this point, it seems fair to say that the unique point of consensus between all the results already presented in the literature is the existence of the p-wave resonances below 1 eV. Bearing in mind that the theoretical works already reported on the theme were carried out with MP [8] and *ab initio* calculations [11,12], a semiempirical approach appears as a natural and convenient alternative to bring light to the problem.

Opposed to *ab initio* and model potential formulations, semiempirical approaches are based on adjustable parameters, usually tuned to reproduce some external previously known quantity. In practice, we are going to work with a single-body potential to represent the electron–atom interaction whose polarization component has asymptotically correct form up to the first order [16,17], and its short range component is adjusted to reproduce the position of the p-wave resonances at the energies observed by Burrow et al. [4].

This paper is organized as follows: in Section 2, we present the exchange and polarization potentials adopted to perform our calculations with a very brief description of the elements of electron–atom scattering theory used to calculate the cross sections; in Section 3, we show and discuss our results with particular attention to the issue of resonances and the contrast with previous data; finally, in Section 4 we synthesize our conclusions. Except where explicitly stated, atomic units are used.

2. Methods and Procedures

Following our previous works [16,18] and particularly inspired by the article of O'Connel and Lane [19], we write the effective Hamiltonian for the electron–atom system as

$$H = -\frac{1}{2}\nabla^2 + V_{st}(\vec{r}) + V_{exc}(\vec{r}) + V_{pol}(\vec{r}). \tag{1}$$

The term $V_{st}(\vec{r})$ represents the electrostatic electron–atom interaction. It directly depends on the model adopted to describe the atomic system and that determines the atomic charge density $\rho(\vec{r})$. Once it is chosen, the calculation of $V_{st}(\vec{r})$ is performed by the usual means [19].

The indistinguishability between the projectile and the target electrons is a non-local effect of expressive computational cost in many-body calculations. Here, we adopt a local model potential $V_{exc}(\vec{r})$ to describe this effect. Several works were performed throughout the history of the electron–atom research field in order to develop model exchange potentials and evaluate their performances [20–23]. Here, we follow Riley and Thrular [22] and chose to work with the Hara free electron gas model (HFEGE) [24] as the recommended model exchange potential for the low-energy domain.

The HFEGE exchange potential is given by

$$V_{exc}(\vec{r}) = -\frac{2}{\pi}K_F(\vec{r})F[\eta(\vec{r})] \tag{2}$$

where the functions $K_F(\vec{r})$, $F(\eta)$, and $\eta(\vec{r})$ have the forms

$$K_F(\vec{r}) = [3\pi\rho(\vec{r})]^{1/3}, \tag{3}$$

$$F(\eta) = \frac{1}{2} + \frac{1-\eta^2}{4\eta}\ln\left|\frac{1+\eta}{1-\eta}\right|, \tag{4}$$

and

$$\eta(\vec{r}) = \frac{K(\vec{r})}{K_F(\vec{r})}, \tag{5}$$

respectively. In these equations, $K_F(\vec{r})$ denotes the Fermi momentum and $\rho(\vec{r})$ is the atomic charge density. $K(\vec{r})$ is the local momentum given by

$$K^2(\vec{r}) = K_F^2(\vec{r}) + 2I + k^2 \tag{6}$$

where I is the ionization potential of the target atom and $k^2/2$ is the incident kinetic energy of the projectile.

The representation of the model exchange interaction demands an accurate representation of the electronic density along the entire space occupied by the atomic target. Contrary to what happens in positron–atom scattering, the incident electron encounters a purely attractive potential which evidently causes it to penetrate more deeply into the target field.

In our previous investigations on positron–atom scattering [15,16,18], we adopted the model electronic density of Salvat et al. [25]. As highlighted by Rabasović et al. [26], the model densities given by Salvat et al. [25] provide a poor representation of the atomic shell structure. In order to overcome this difficulty, we adapted our codes to work with SCF/HF electronic densities. Among many possibilities, we selected the DZP basis as given in references [27,28] for Zn and Cd, respectively, discarding F type functions, and constructed the SCF/HF ground state wave functions with the GAMESS package [29].

Figure 1 shows the radial density profile (RDP) calculated with the two prescriptions for comparison. The atomic shell structure is evidently much better represented by the SCF model. We also visualize that for $r > 2.5\ a_0$ the RDPs converge among each other for both atoms. This point explains why the use of model densities such as the one of Salvat et al. [25] usually poses no problems in the positron case; since the positron–atom static potential is repulsive and exchange is not present, the positron is essentially scattered at the edge of the target. In practice, any model potential will depend only on the values of the atomic charge density in the peripheral region of the system.

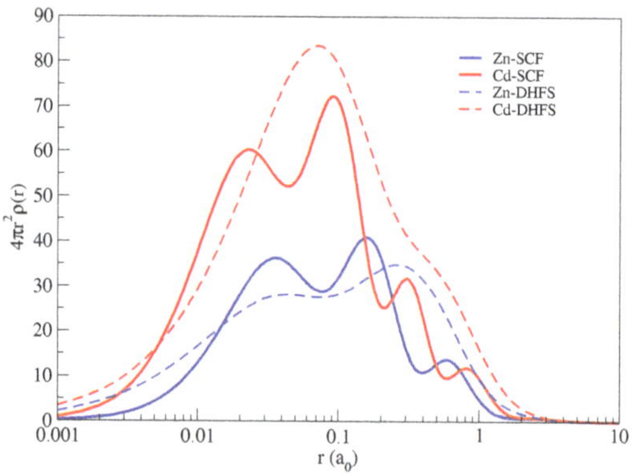

Figure 1. Radial probability density for Zn and Cd atoms obtained from Hartree–Fock (SCF) wavefunctions (this work) for Zn and Cd compared to the ones calculated with model atomic charge densities obtained from the fitting of Dirac–Hartree–Fock–Slater (DHFS) calculations as given by Salvat et al. [25].

Finally, the last of the interaction terms is the polarization potential $V_{pol}(\vec{r})$, whose asymptotic form is, up to the first order of perturbation theory, given by

$$\lim_{r\to\infty} V_{pol}(r) \approx -\frac{\alpha_d}{2r^4}, \tag{7}$$

where α_d is the static dipole polarizability [30]. Here, we adopt the same semiempirical potential as considered in Arretche et al. [18]:

$$V_{pol}(r;r_c) = -\frac{\alpha_d}{2r^4}\left[1 - e^{-(r/r_c)^6}\right] \tag{8}$$

where r_c is an adjustable parameter. Since the experimental data reported by Burrow et al. [4] are essentially a measure of the transmitted current (and no partial wave analysis is performed), we decided to assign the values of r_c where the peak of the elastic cross section computed with all partial waves matches the energy position of the p-wave shape resonances. Coincidentally, for both the systems we have found $r_c = 3.400\ a_0$ as the value that satisfies this criterion.

Table 1 shows the values of the static dipole polarizabilities α_d for each target and the respective cutoff radius r_c that reproduces the position of the p-wave shape resonances at the energies (E (eV)) reported by Burrow et al. [4]. The values for α_d were taken from our recent investigation on low-energy positron scattering by Zn and Cd [15] where the influence of higher order polarizabilities in positron–Zn/Cd was explored. For the sake of completeness, the ionization potentials (I (eV)) [31] are also included.

Table 1. Values adopted for the static dipole polarizabilities α_d (in a_0^3) for each target and the respective cutoff radius r_c (in a_0) that reproduces the position (E (eV)) of the p-wave shape resonances at the energies reported by Burrow et al. [4]. I (eV) denotes the ionization potentials for each target [31].

Atom	α_d	E (eV)	I (eV)
Zn	39.20	0.49	9.39
Cd	45.92	0.33	8.99

In Figure 2, we show the scattering potentials for reference. As expected, the potentials for Zn and Cd are rather alike since they exhibit very similar electronic structures. Once the potentials were defined, the cross sections were computed as in Arretche et al. [18].

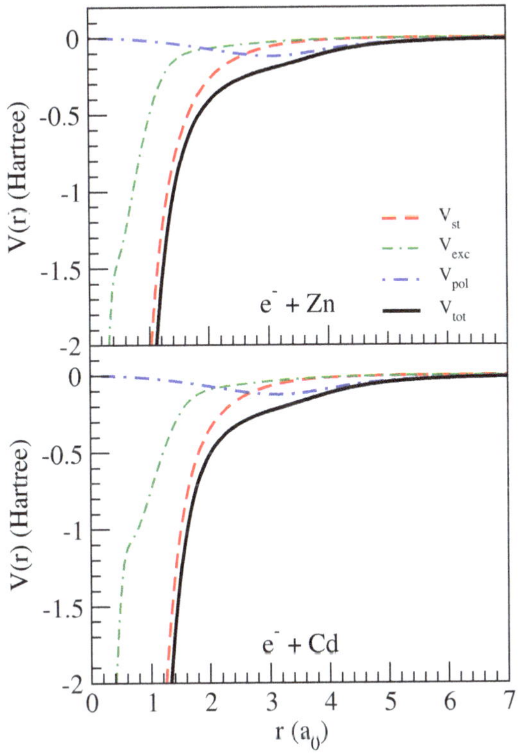

Figure 2. Electron–Zn (**upper**) and electron–Cd (**bottom**) scattering potentials. Legends are in the figure.

3. Results and Discussion

Unquestionably, measurements of the differential cross sections (DCSs) are a great challenge in the area of electron/positron–atom scattering, mainly in the very low-energy region. It is not surprising to find a single measurement of relative elastic DCSs made by Marinković for electron–Cd at 3.4 eV [5]. While TCS measurements based on the Beer–Lambert law are the basic test of any theory, it is the DCS that translates the fine tuning between different components of the scattering potential. For example, in positron–molecule scattering, it is the degree of polarization considered in the potential which affects the position of the minima at the intermediate angles in the elastic DCS while the scattering at high angles (backward scattering) is regulated by the repulsive short range static potential [32].

The relative elastic DCS reported by Marinković et al. [5] at 3.4 eV for Cd is shown in Figure 3 where the CCC and RCCC of Berrington et al. [12] are also shown for comparison. Except for the region between 90 and 120 degrees, the three theoretical DCSs are very close to each other. The DCS of Marinković et al. [5] was normalized at 50 degrees where CCC, RCCC, and our results find the maximum agreement. Interestingly, our results are very similar to the RCCC ones. On the other hand, the CCC DCS seems to better describe the experimental data between 90 and 150 degrees. At present, the lack of data, theoretical and experimental, does not allow us to go any further in the analysis of this aspect of the problem.

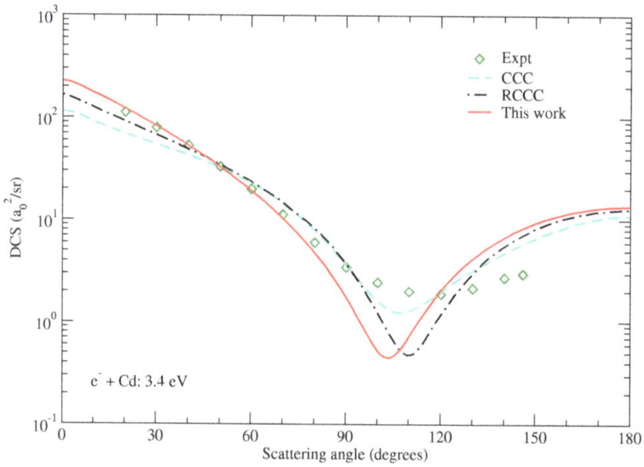

Figure 3. Differential elastic cross section for electron–Cd at 3.4 eV compared to the data of Marinković et al. [5] and Berrington et al. [12]. The relative DCS data were normalized at 50 degrees.

We then pay attention to the elastic integral cross section (ICS) for Zn and Cd which are given in the upper and bottom panels of Figure 4, respectively. In both plots, the cross sections computed with our semiempirical approach are shown as a solid red line.

In a given sense, our calculations are somewhat similar to the MP results presented by Yuan and Zhang [8] since both treat the correlation–polarization effect by single-body potentials. In reference [8], two sets of cross sections were reported for each target. The first set was obtained from the PZ correlation potential while the second one was generated by scaling the correlation potential as $V_c \rightarrow \alpha V_c$ with α chosen to "remove the overestimation of the correlation effect" and reproduce the ^2P-shape resonance of e-Mg scattering at 0.15 eV [33]. In [33], α was taken as 0.8 and the same value was also applied for Zn and Cd but without any particular justification or connection to the resonance data of Burrow et al. [4]. Due to that, we have chosen to show only the unscaled cross sections. We can clearly see in Figure 4 that for both atoms, the MP cross sections (given as a black dashed line) shift the p-wave resonances to lower energies than the ones experimentally observed (and reproduced by us) and with narrower widths.

We also consider the results provided by the *ab initio* many-body techniques. For Zn, we show the core potential (CPRM) and the B-spline R-matrix (BSRM) cross sections of Zatsarinny and Bartschat [11]. The core potential approach, as the name suggests, is based, roughly speaking, on the idea of representing the Hartree potential of the Zn^{2+} by semiempirical exchange and polarization potentials while the valence "helium-like" region is treated *ab initio*. The B-spline technique, on the other hand, uses the B-splines as a universal basis to represent the electron scattering orbitals. The p-wave resonance energy predicted by the BSRM calculation occurs at 0.707 eV, a little bit higher than the experimental one with the CPRM result at an intermediate energy. MP, CPRM, and BSRM furnish similar cross sections above $\approx$1 eV while our semiempirical calculations show another resonance at $\approx$2.83 eV. The elastic integral cross section below 4 eV is not explicitly given by Berrington et al. [12], as the only direct information that can be extracted about it comes from the DCS at 3.4 eV (see Figure 6 of [12]). The ICS calculated from it is given as a square in the bottom panel. As in Zn, above $\approx$1 eV the MP, RCCC, and our semiempirical cross sections perfectly agree with each other. On the other hand, notable discrepancy is found between the theoretical results and the experimental data of Kontros et al. [6]. Except for a scale factor, our semiempirical ICS presents the same qualitative dependence with energy when compared to the experimental points, from threshold up to $\approx$0.5 eV, with the p-wave resonance observed by Burrow et al. [4] being found approximately at the same energy (0.33 eV). Notwithstanding, above 0.5 eV notable divergence is evident.

In order to better appreciate the resonant structures, we show in Figure 5 the partial wave cross sections found in our calculations. As stated in the Introduction, the possible existence of a d-wave shape resonance for both Zn and Cd between the p-wave shape resonance and the first excited states of the neutral atom was suggested by Burrow et al. [4]. Our results show a d-wave resonance for Zn at 2.83 eV but a flat d-wave cross section for Cd. Zatsarinny and Bartschat [11] have also found a d-wave resonance with a BSRM at 4.234 eV (see Table III of [11]). Since shape resonances are basically characterized by their energies and widths, we present these values in Table 2.

Finally, for the sake of completeness, we report the scattering lengths obtained in our calculations. These values are 2.32 a_0 for Zn and 2.24 a_0 for Cd.

Table 2. Resonance widths in eV. The experimental values are the ones of Burrow et al. [4]; BSRM is the B-spline R-matrix of Zatsarinny and Bartschat [11].

Atom		Expt	BSRM	This Work
Zn	p	0.45	1.14	0.95
	d		0.37	0.70
Cd	p	0.33		0.55

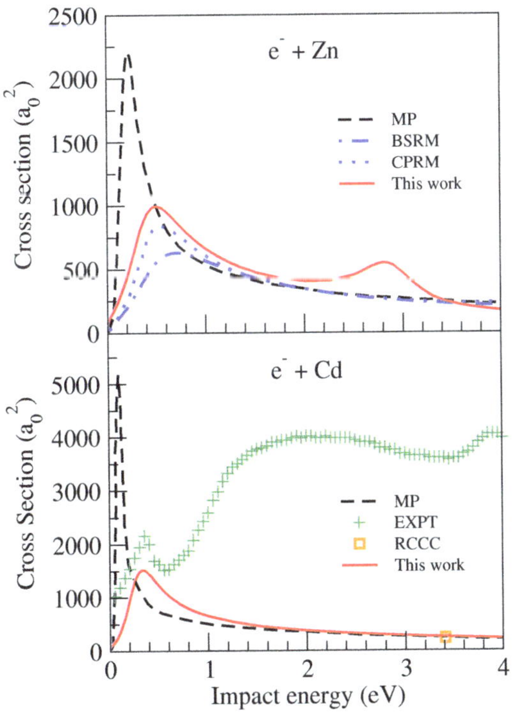

Figure 4. Elastic cross section for electron–Zn (**upper** panel) and electron–Cd (**bottom** panel). In both panels, the present results are given by the solid red line, and the MP calculation results of [8] are represented by the dashed black line. For Zn, the dotted blue line represents the CPRM calculation and the dashed blue line the BSRM results [11]. For Cd, the green crosses are the experimental data of Kontros et al. [6], and the yellow square is the ICS obtained by integration of the RCCC [12] DCS at 3.4 eV.

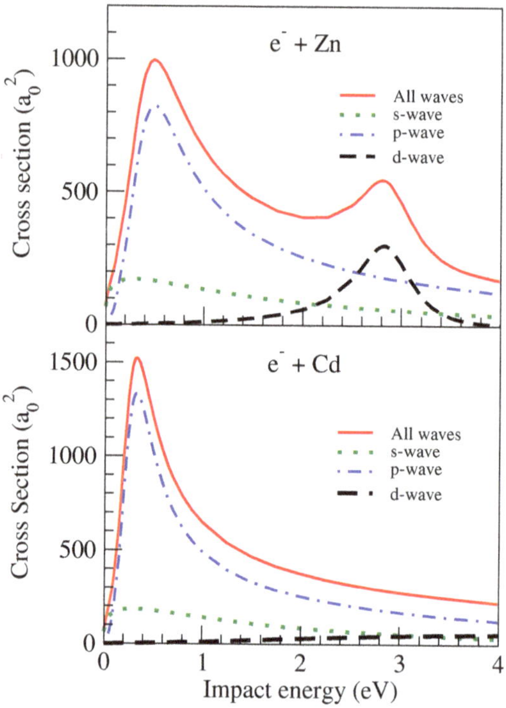

Figure 5. Partial wave contributions to the elastic cross section for electron–Zn (**upper** panel) and electron–Cd (**bottom** panel).

4. Conclusions

The low-energy scattering of electron–Zn/Cd was studied by applying model exchange and semiempirical polarization potentials. The external data used to adjust the short range component of the polarization potential were the resonance data of Burrow et al. [4]. Given this methodology, we reported the elastic integral cross sections for energies below 4 eV and the scattering length for both targets.

Our results for the Cd DCS at 3.4 eV are very close to the RCCC of Berrington et al. [12], but showed small discrepancies with the CCC [12] and the experimental DCS of Marinković et al. [5] in the intermediate angular region. The elastic ICSs for Zn are in good agreement with the previous MP and *ab initio* results but contrasted with the previous ones, and we have found a clear d-wave resonance as conjectured by Burrow et al. [4] and pointed by Clark and Buckman [14]. For Cd, no d-wave resonance was found, but our cross sections corroborate the previous ones obtained with MP and *ab initio* (only for 3.4 eV). Even so, all of the theoretical cross sections have a great discrepancy with the experimental data of Kontros et al. [6].

The calculations presented here show that even for relatively well-documented atomic systems such as metal vapors, there are still several gaps to be filled. It would be interesting to see a set of low-energy total cross section measurements for Zn and Cd in order to contrast with the data of Burrow et al. [4] and, more specifically, with the measurements of Kontros et al. [6] since these are in disagreement with the theoretical predictions.

Author Contributions: Conceptualization, F.A., W.T. and B.K.S.; methodology, F.A. and W.T.; writing—original draft preparation, F.A. and W.T.; writing—review and editing, F.A., W.T. and B.K.S. All authors have read and agreed to the published version of the manuscript.

Funding: This research received no external funding.

Institutional Review Board Statement: Not applicable.

Informed Consent Statement: Not applicable.

Data Availability Statement: The associated data will be available from authors upon request.

Acknowledgments: F. Arretche and W. Tenfen would like to thank the Programa de Pós-Graduação em Física of Universidade Federal de Santa Catarina and Universidade Federal de Pelotas for the support. B.K.S. acknowledges the use of the Vikram-100 HPC facility of PRL, India.

Conflicts of Interest: The authors declare no conflict of interest.

Abbreviations

The following abbreviations are used in this manuscript:

DCS	Differential cross section
CCC	Convergent close-coupling
RCCC	Relativistic convergent close-coupling
MP	Model potential
HFEGE	Hara free electron gas
SCF/HF	Self consistent field/Hartree–Fock
GAMESS	General Atomic and Molecular Electronic Structure System
TCS	Total cross section
ICS	Integral cross section
CPRM	Core potential R-matrix
BSRM	B-spline R-matrix

References

1. Hopwood, J.A. Plasma physics. In *Thin Films*; Elsevier: Amsterdam, The Netherlands, 2000; Volume 27, pp. 181–207.
2. Childs, E.C.; Massey, H.S.W.; Rutherford, E. The scattering of electrons by metal vapours. I—Cadmium. *Proc. R. Soc. Lond. Ser. A Contain Pap. A Math. Phys. Charact.* **1933**, *141*, 473–483. [CrossRef]
3. Childs, E.C.; Massey, H.S.W. The scattering of electrons by metal vapours. II—Zinc. *Proc. R. Soc. Lond. Ser. A Contain Pap. A Math. Phys. Charact.* **1933**, *142*, 509–518. [CrossRef]
4. Burrow, P.; Michejda, J.; Comer, J. Low-energy electron scattering from Mg, Zn, Cd and Hg: Shape resonances and electron affinities. *J. Phys. B At. Mol. Phys.* **1976**, *9*, 3225. [CrossRef]
5. Marinkovic, B.; Pejcev, V.; Filipovic, D.; Vuskovic, L. Elastic and inelastic electron scattering by cadmium. *J. Phys. B At. Mol. Opt. Phys.* **1991**, *24*, 1817–1837. [CrossRef]
6. Kontros, J.; Szoter, L.; Chernyshova, I.; Shpenik, O. Cross sections of slow electron scattering by cadmium atoms. *J. Phys. B At. Mol. Opt. Phys.* **2002**, *35*, 2195. [CrossRef]
7. Sullivan, J.; Burrow, P.; Newman, D.S.; Bartschat, K.; Michejda, J.; Panajotovic, R.; Moghbelalhossein, M.; McEachran, R.; Buckman, S. An experimental and theoretical study of transient negative ions in Mg, Zn, Cd and Hg. *New J. Phys.* **2003**, *5*, 159. [CrossRef]
8. Yuan, J.; Zhang, Z. Elastic cross sections of electron scattering with Zn and Cd atoms below 4 eV. *Phys. Rev. A* **1992**, *45*, 535. [CrossRef]
9. Perdew, J.P.; Zunger, A. Self-interaction correction to density-functional approximations for many-electron systems. *Phys. Rev. B* **1981**, *23*, 5048. [CrossRef]
10. Padial, N.T.; Norcross, D.W. Parameter-free model of the correlation-polarization potential for electron-molecule collisions. *Phys. Rev. A* **1984**, *29*, 1742–1748. [CrossRef]
11. Zatsarinny, O.; Bartschat, K. Benchmark calculations for electron collisions with zinc atoms. *Phys. Rev. A* **2005**, *71*, 022716. [CrossRef]
12. Berrington, M.J.; Bostock, C.J.; Fursa, D.V.; Bray, I.; McEachran, R.; Stauffer, A. Calculations of electron scattering from cadmium. *Phys. Rev. A* **2012**, *85*, 042708. [CrossRef]
13. McEachran, R.; Marinković, B.; García, G.; White, R.; Stokes, P.; Jones, D.; Brunger, M. Integral cross sections for electron–zinc scattering over a broad energy range (0.01–5000 eV). *J. Phys. Chem. Ref. Data* **2020**, *49*, 013102. [CrossRef]
14. Buckman, S.J.; Clark, C.W. Atomic negative-ion resonances. *Rev. Mod. Phys.* **1994**, *66*, 539. [CrossRef]
15. Arretche, F.; Andermann, A.M.; Seidel, E.P.; Tenfen, W.; Sahoo, B.K. Polarization effects, shape resonances and bound states in low energy positron elastic scattering by Zinc and Cadmium vapours. *J. Electron Spectrosc. Relat. Phenom.* **2022**, *257*, 147186. [CrossRef]
16. Arretche, F.; Barp, M.V.; Scheidt, A.; Seidel, E.P.; Tenfen, W. Semiempirical models for low energy positron scattering by Ar, Kr and Xe. *J. Phys. B At. Mol. Opt. Phys.* **2019**, *52*, 215201. [CrossRef]

17. Tenfen, W.; Seidel, E.P.; Barp, M.V.; Arretche, F. Higher order polarizabilities and the positron forward scattering problem: Convergence between calculated and measured cross sections in the very low energy regime. *J. Electron Spectrosc. Relat. Phenom.* **2022**, *257*, 147160. [CrossRef]
18. Arretche, F.; Barp, M.V.; Tenfen, W.; Seidel, E.P. The Hidden Ramsauer-Townsend Effect in Positron Scattering by Rare Gas Atoms. *Braz. J. Phys.* **2020**, *50*, 844–856. [CrossRef]
19. O'Connell, J.K.; Lane, N.F. Nonadjustable exchange-correlation model for electron scattering from closed-shell atoms and molecules. *Phys. Rev. A* **1983**, *27*, 1893–1903. [CrossRef]
20. Furness, J.; McCarthy, I. Semiphenomenological optical model for electron scattering on atoms. *J. Phys. B At. Mol. Phys.* **1973**, *6*, 2280. [CrossRef]
21. Bransden, B.; McDowell, M.; Noble, C.; Scott, T. Equivalent exchange potentials in electron scattering. *J. Phys. B At. Mol. Phys.* **1976**, *9*, 1301. [CrossRef]
22. Riley, M.E.; Truhlar, D.G. Approximations for the exchange potential in electron scattering. *J. Chem. Phys.* **1975**, *63*, 2182–2191. [CrossRef]
23. Yau, A. Electron scattering from noble gases. *J. Phys. B At. Mol. Phys.* **1978**, *11*, 2907. [CrossRef]
24. Hara, S. The scattering of slow electrons by hydrogen molecules. *J. Phys. Soc. Jpn.* **1967**, *22*, 710–718. [CrossRef]
25. Salvat, F.; Martínez, J.D.; Mayol, R.; Parellada, J. Analytical Dirac-Hartree-Fock-Slater screening function for atoms (Z=1–92). *Phys. Rev. A* **1987**, *36*, 467–474. [CrossRef] [PubMed]
26. Rabasović, M.; Kelemen, V.; Tošić, S.; Šević, D.; Dovhanych, M.; Pejčev, V.; Filipović, D.; Remeta, E.Y.; Marinković, B. Experimental and theoretical study of the elastic-electron–indium-atom scattering in the intermediate energy range. *Phys. Rev. A* **2008**, *77*, 062713. [CrossRef]
27. Camiletti, G.; Machado, S.; Jorge, F. Gaussian basis set of double zeta quality for atoms K through Kr: Application in DFT calculations of molecular properties. *J. Comput. Chem.* **2008**, *29*, 2434–2444. [CrossRef]
28. Barros, C.; De Oliveira, P.; Jorge, F.; Canal Neto, A.; Campos, M. Gaussian basis set of double zeta quality for atoms Rb through Xe: application in non-relativistic and relativistic calculations of atomic and molecular properties. *Mol. Phys.* **2010**, *108*, 1965–1972. [CrossRef]
29. Barca, G.M.J.; Bertoni, C.; Carrington, L.; Datta, D.; De Silva, N.; Deustua, J.E.; Fedorov, D.G.; Gour, J.R.; Gunina, A.O.; Guidez, E.; et al. Recent developments in the general atomic and molecular electronic structure system. *J. Chem. Phys.* **2020**, *152*, 154102. [CrossRef]
30. Bransden, B.; Joachain, C.; Plivier, T. *Physics of Atoms and Molecules*; Pearson Education, Prentice Hall: Hoboken, NJ, USA, 2003.
31. Iliaš, M.; Neogrády, P. Ionization potentials of Zn, Cd, Hg and dipole polarizabilities of Zn^+, Cd^+, Hg^+: Correlation and relativistic effects. *Chem. Phys. Lett.* **1999**, *309*, 441–449. [CrossRef]
32. Tenfen, W.; Barp, M.V.; Arretche, F. Low-energy elastic scattering of positrons by O_2. *Phys. Rev. A* **2019**, *99*, 022703. [CrossRef]
33. Yuan, J.; Zhang, Z. The low-lying shape resonances in low-energy electron scattering with Be, Mg and Ca atoms. *J. Phys. B At. Mol. Opt. Phys.* **1989**, *22*, 2751. [CrossRef]

atoms

MDPI

Article

Inclusion of Electron Interactions by Rate Equations in Chemical Models

Laurence Campbell [1,*], Dale L. Muccignat [2] and Michael J. Brunger [1,3]

[1] College of Science and Engineering, Flinders University, Bedford Park, SA 5042, Australia; michael.brunger@flinders.edu.au
[2] College of Science and Engineering, James Cook University, Townsville, QLD 4810, Australia; dale.muccignat@my.jcu.edu.au
[3] Institute of Actuarial Science and Data Analytics, UCSI University, Kuala Lumpur 56000, Malaysia
* Correspondence: laurence.campbell@flinders.edu.au

Abstract: The concept of treating subranges of the electron energy spectrum as species in chemical models is investigated. This is intended to facilitate simple modification of chemical models by incorporating the electron interactions as additional rate equations. It is anticipated that this embedding of fine details of the energy dependence of the electron interactions into rate equations will yield an improvement in computational efficiency compared to other methods. It will be applicable in situations where the electron density is low enough that the electron interactions with chemical species are significant compared to electron–electron interactions. A target application is the simulation of electron processes in the D-region of the Earth's atmosphere, but it is anticipated that the method would be useful in other areas, including enhancement of Monte Carlo simulation of electron–liquid interactions and simulations of chemical reactions and radical generation induced by electrons and positrons in biomolecular systems. The aim here is to investigate the accuracy and practicality of the method. In particular, energy must be conserved, while the number of subranges should be small to reduce computation time and their distribution should be logarithmic in order to represent processes over a wide range of electron energies. The method is applied here to the interaction by inelastic and superelastic collisions of electrons with a gas of molecules with only one excited vibrational level. While this is unphysical, it allows the method to be validated by checking for accuracy, energy conservation, maintenance of equilibrium and evolution of a Maxwellian electron spectrum.

Keywords: electron scattering; chemical model; simulation; rate equations; electron energy distribution

Citation: Campbell, L.; Muccignat, D.L.; Brunger, M.J. Inclusion of Electron Interactions by Rate Equations in Chemical Models. *Atoms* **2022**, *10*, 62. https://doi.org/10.3390/atoms10020062

Academic Editors: Rajesh Srivastava and Dmitry V. Fursa

Received: 20 May 2022
Accepted: 8 June 2022
Published: 10 June 2022

1. Introduction

Simulations of atmospheric processes involving chemistry often involve the processing of a set of reactions specified in the form:

$$\mathrm{A} + \mathrm{B} \xrightarrow{k} \mathrm{C} + \mathrm{D} + \ldots\,, \tag{1}$$

where A, B, C and D are atoms or molecules and k is the rate constant. The rate r at which A and B interact is given by:

$$r = k[\mathrm{A}][\mathrm{B}]\,, \tag{2}$$

where $[x]$ represents the density of species x and the rate constant k has the units $\mathrm{L}^3\mathrm{s}^{-1}$ where "L" is the unit of length in which the density is specified (e.g., $\mathrm{cm}^3\mathrm{s}^{-1}$).

For each species i with number density n_i, a one-dimensional mass continuity equation can be written:

$$\frac{\delta n_i}{\delta t} = P_i - l_i n_i\,, \tag{3}$$

where P_i and l_i are the production rate and loss probability [1].

The set of continuity equations can be solved iteratively until equilibrium is obtained [2], or applied over a series of time steps for non-equilibrium simulations [1]. Even where equilibrium is required, a time-step simulation is useful to determine the time for equilibrium to be attained [3].

Electron interactions can be (e.g., [4]) incorporated into the set of rate equations in forms such as:

$$A^+ + e \xrightarrow{k_r} A \text{ or } A + e \xrightarrow{k_a} A^- \text{ or } A + e \xrightarrow{k_{ex}} A^* + e^* , \qquad (4)$$

where * represents a change in energy and k_r, k_a and k_{ex} are the average rate constants, for recombination, attachment and excitation, for electrons with a Maxwellian distribution characterised by an electron temperature T_e. However, in many atmospheric applications, such as those including auroral, cosmic-ray or VUV input, the electrons are not in thermal equilibrium [5,6]. For example, in the Earth's nighttime mesosphere, electrons are created in ionization produced by cosmic rays and by Lyman-α radiation [7], and they then lose energy in collisions with atmospheric molecules before recombining with positive ions or attaching to molecules to produce negative ions. At the same time, chemical processes produce vibrationally-excited OH [8], with some of this energy being transferred to other species and then to free electrons in superelastic collisions. The time-scale of these processes depends on the density of each constituent. Therefore, calculating the electron density requires a non-equilibrium simulation that includes both the electron–impact processes and chemical processes.

Another potential application may be to enhance Monte Carlo simulation of non-equilibrium electron–liquid transport [9]. In cases where there are too many tracks (from ionisation) or too few tracks (from attachment) to simulate, approximate approaches such as "re-scaling" [10] or "weighting" [11] are used and the distribution of the excited species is not considered. It is possible that using a chemical model that includes electron interactions could be applied in such situations. In addition, the chemical model could be applied to determine a non-equilibrium distribution of excited species in a gas prior to applying a Monte Carlo simulation of electron impact.

Ionizing radiation applied to biological materials can produce secondary electrons, radicals (e.g., $OH^\bullet$) and ions (e.g., H_3O^+), all of which can cause damage to DNA [12,13]. As the electrons may lose energy in a series of interactions with water and other molecules before interacting with a DNA molecule, a chemical model that incorporates a changing electron energy spectrum should be applicable to simulating damage to DNA by ionizing radiation.

Thus, an aim of this work is to develop a method to calculate the development of the electron energy spectrum for a system where the electrons are interacting with atoms and molecules that are simultaneously interacting with each other. This has previously been addressed in plasma physics as part of a "state-to-state" approach [14] in which excited states of atoms and molecules are treated as independent species. This chemical model is coupled to the electron distribution via the Boltzmann equation, with the reaction rates between electrons and chemical species being recalculated each time the coupling is made. The aim here is to do this with a different approach in which groups of electrons of similar energy are treated as individual chemical species and consequently all interactions are calculated in each time step. An advantage of this method is that the electron interactions can be added into the list of rate equations in existing models, avoiding the work involved in merging different computer codes that are geared to different problems. Another advantage is that some detail (e.g., thresholds and narrow resonances) of the electron interactions can be incorporated into the reaction rates, so they do not have to be recalculated at each time step and so a relatively coarse energy grid can be used to reduce computation time.

The proposed method is to divide the electron energy range into subranges and to treat each subrange in the same way as a chemical species i.e., each reaction, such as those in Equation (4), is replaced by a series of reactions that represent the interaction for a set of different initial and final electron–energy subranges. Due to the large range of energies of different processes (from ionisation to recombination), a logarithmic distribution of the energy ranges is preferable.

A test of this proposed method, to investigate validity and accuracy, is to simulate the injection of electrons of one energy into a gas of OH molecules, where only vibrational excitation of one level of OH is considered. OH is chosen as it is relevant to two of the examples postulated above.

In Section 2.1, the background theory and computational techniques required to develop and test the proposed method are outlined. In Section 2.2, the method and its computational implementation are outlined. The results of various tests of the method to show its viability and the errors involved are given in Section 3. These results are discussed in Section 4, from which conclusions are drawn and given in Section 5.

2. Materials and Methods

2.1. Background Theory and Techniques

A simple way to simulate the evolution of a set of interacting species is to apply Equation (2) to each reaction (1) for a time-step Δt, so that the change in each species is:

$$-\Delta[\mathrm{A}] = -\Delta[\mathrm{B}] = \Delta[\mathrm{C}] = \Delta[\mathrm{D}] = k[\mathrm{A}][\mathrm{B}]\Delta t. \quad (5)$$

For each species i, the gains and losses are added up to give the total gain G_i and total loss L_i so the new density n_i of species i after time Δt is:

$$n_i(t+\Delta t) = n_i(t) + G_i - L_i. \quad (6)$$

The development of the densities of all species can be simulated by iterative application of Equation (6) over the required time, but the magnitude of Δt is limited by the requirement that the density must not go negative and so this "explicit" formula is unsuitable for simulation of systems over long time intervals.

An alternative "semi-implicit" method [15] is justified by the fact that, through the time interval Δt, the loss rate l_i is proportional to the instantaneous density $n_i(t)$. Hence, an approximation is being made irrespective of whether the value of L_i is approximated by being proportional to $n_i(t)$ or $n_i(t+\Delta t)$. Thus, substituting the final density for the initial density:

$$n_i(t+\Delta t) = n_i(t) + G_i - L_i = n_i(t) + G_i - L_i\frac{n_i(t)}{n_i(t)} = n_i(t) + G_i - L_i\frac{n_i(t+\Delta t)}{n_i(t)} \quad (7)$$

and rearranging:

$$n_i(t+\Delta t)\left(1 + L_i\frac{1}{n_i(t)}\right) = n_i(t) + G_i\,, \quad (8)$$

leads to:

$$n_i(t+\Delta t) = \frac{n_i(t) + G_i}{1 + \frac{L_i}{n_i(t)}}\,. \quad (9)$$

As the new density cannot be negative, Equation (9) avoids the main problem with Equation (6) and so allows much longer values of Δt to be used.

There are more sophisticated time-step algorithms (e.g., [15,16]), but the requirement here is to be able to discriminate between the error due to the time-step algorithm and any error inherent in the approximation of the electron energy spectrum. For example, the Gauss–Seidel method is quoted to have an error of 1% [16]. As Equation (9) would be expected to have no error in the limit of very small time steps, the errors due to the simulation method can be separated from the errors due to the time-step calculation.

To allow the simulation to run to equilibrium with minimum computation time, an adaptive time step Δt is required to implement Equation (9) efficiently [3]. The initial value of Δt is set very small (e.g., 10^{-10} s) and is then successively increased as:

$$\Delta t_{j+1} = f\Delta t_{\min} = f \min\left[\frac{n_i(t)}{L_i - G_i} \text{ for } n_i(t) > 1, L_i > G_i\right]\Delta t_j\,, \tag{10}$$

where $\Delta t_{\min}$ is the minimum time interval for any of the constituents i to fall to zero in the next time step Δt_{j+1} at the current rate of change, and f is a fraction that acts to reduce the error in the calculations.

The rate constant for collisions between an electron and a gas molecule in a unit volume is $v\sigma$, where v is the electron's velocity and σ is the cross section or probability for the interaction. Thus, the rate constant as a function of energy is:

$$k(E) = v\sigma = \sigma\sqrt{2E/m}. \tag{11}$$

It therefore follows that the rate constant $k(E_a, E_b)$ for all transitions of electrons starting in a range $[E_a, E_b]$ is:

$$k(E_a, E_b) = \frac{\int_{E_a}^{E_b} \sigma(E)\sqrt{2E/m}F(E)dE}{\int_{E_a}^{E_b} F(E)dE}\,, \tag{12}$$

where $F(E)$ is the electron energy distribution function (EEDF). For a Maxwell–Boltzmann distribution:

$$F(E) = \frac{2}{\sqrt{\pi}}\left(\frac{1}{k_B T}\right)^{3/2} E^{1/2}e^{-E/(k_B T)}\,, \tag{13}$$

where T is the electron temperature, and k_B is Boltzmann's constant [17].

The cross section σ_s for a superelastic collision, for electron impact energy E, can be determined from the inelastic cross section using [18,19]:

$$\sigma_s(E) = \frac{E+\varepsilon}{E}\sigma(E+\varepsilon)\,, \tag{14}$$

where ε is the threshold energy of the inelastic excitation.

Electrons impacting atoms and molecules will gain or lose energy by elastic scattering. Published cross sections [20] for elastic electron scattering by OH are only for electron energy above 1 eV. Thus, as an approximation, the formula for the elastic electron energy transfer rate for O of Banks [21] is used, i.e.,

$$dU_e/dt = -3.74 \times 10^{-18} n_e n(\mathrm{O}) T_e^{1/2}(T_e - T) \tag{15}$$

where dU_e/dt is the energy transfer rate (eV $\mathrm{cm}^{-3}\mathrm{s}^{-1}$), T_e is the temperature of a Maxwellian distribution of electrons, T is the temperature of the O atoms, n_e is the electron density (cm^{-3}), and $n(\mathrm{O})$ is the O density.

2.2. Proposed Method

To incorporate the electron reactions (4) into a time-step simulation, the proposed approach is to split the electron energy range $[E_{\min}, E_{\max}]$ into N subranges R_1–R_N, with R_0 set as the range $[0, E_{\min}]$.

In place of the single reaction $A + e \xrightarrow{k} A^* + e^*$, a series of reactions:

$$A + R_i \xrightarrow{k_{ij}} A^* + R_j\,, \tag{16}$$

is entered into the list of chemical reactions, with the number density of electrons in each energy range R_i treated in the same way as the density of a chemical species. (For example, in the current implementation, the number of electrons in the range $[0, E_{min}]$ is stored in a variable R_0. While a logarithmic distribution of subranges is desirable, a linear distribution is considered first for simplicity before proceeding to the complications added by a logarithmic distribution.

Consider a case where electrons lose or gain energy ε in elastic and superelastic collisions with gas molecules. In Figure 1a, the case where the transition energy ε is less than the size of the energy subranges is considered for two energy subranges with boundaries at E_{01}, E_{12} and E_{23} and midpoints at E_1 and E_2. Applying Equation (12) with $F(E) = 1$ (as the electron–energy distribution is varying), the rate constant k_{21} for transitions where the electron crosses from R_2 to R_1 is:

$$k_{21} = \frac{\int_{E_{12}}^{E_{12}+\varepsilon} \sigma(E)\sqrt{2E/m}dE}{\int_{E_{12}}^{E_{23}} dE} \tag{17}$$

The rate constant k_{22}, for cases where the electron energy remains in the same subrange while the excited species is produced is similarly:

$$k_{22} = \frac{\int_{E_{12}+\varepsilon}^{E_{23}} \sigma(E)\sqrt{2E/m}dE}{\int_{E_{12}}^{E_{23}} dE} \tag{18}$$

If $\sigma(E)\sqrt{2E/m}$ were constant, then energy would be conserved because the energy lost by electrons that remain in the same subrange would be offset by the higher energy implied for those that transition to the next subrange, i.e.,

$$k_{21}(E_2 - E_1) = (k_{21} + k_{22})\varepsilon. \tag{19}$$

However, as $\sigma(E)\sqrt{2E/m}$ varies across the subrange, energy will not be conserved so a correction is necessary. This is made by solving Equation (19) for a modified value of k_{22}:

$$k'_{22} = k_{21}(E_2 - E_1 - \varepsilon)/\varepsilon. \tag{20}$$

In the case of Figure 1b, where ε is greater than the subrange size, all electrons transition to a lower subrange and there is no physical value of k_{33}. However, a notional value of k'_{33} can be calculated to maintain conservation of energy:

$$k'_{33} = (k_{32}(E_3 - E_2 - \varepsilon) + k_{31}(E_3 - E_1 - \varepsilon))/\varepsilon. \tag{21}$$

While k'_{33} is an unphysical quantity and can be negative, it implements conservation of energy by correcting for the approximation that $\sigma(E)\sqrt{2E/m}$ is constant over the subrange. Applying a similar analysis for the case where ε is much larger than the subrange size leads to a general equation:

$$k'_{ll} = \Sigma_{i=0}^{l-1} k_{li}(E_l - E_i - \varepsilon)/\varepsilon \tag{22}$$

for inelastic collisions and

$$k'_{ll} = \Sigma_{i=l+1}^{N} k_{li}(E_i - E_l - \varepsilon)/\varepsilon \tag{23}$$

for superelastic collisions.

As electrons may be initially introduced at high energy, but then proceed to lose energy through a series of collision processes down to a very low energy, it is desirable to make the energy subrange boundaries on a logarithmic scale to keep the total number of subranges $N + 1$ to a viable minimum for the calculations while allowing for adequate resolution at low electron energies. Equations (22) and (23) are applicable, even where electrons in one subrange can transition to several lower subranges.

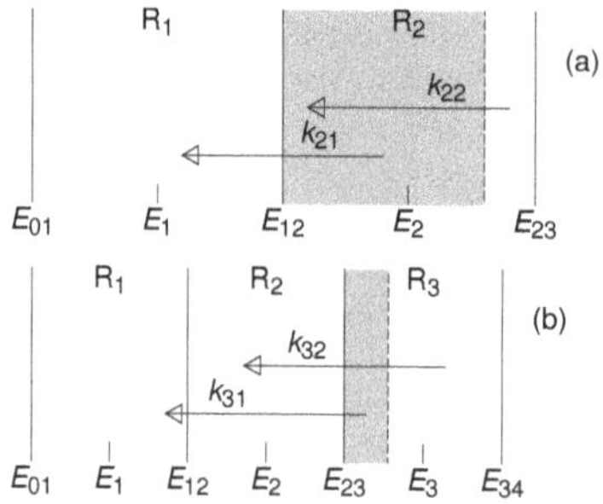

Figure 1. Electron energy losses (arrows) in relation to electron–energy subranges: (**a**) For an energy transition ε that is less than the subrange size $E_{23} - E_{12}$, electrons with initial energies in the shaded area cross the boundary to range R_1, while the others remain within range R_2. (**b**) For ε greater than the subrange size $E_{34} - E_{23}$, all electrons transit to a lower subrange, with those in the shaded region going to R_1 and the others to R_2.

Aspects of the model are illustrated in Figure 2, for inelastic and superelastic collisions of electrons with OH molecules in the lowest and first vibrational level of the ground state. It shows rate constants for individual energies and average rate constants for the simulation where the electron energy range is divided into 30 logarithmically-spaced energy subranges between 0.01 eV and 20 eV, labelled R_1–R_{30}, with electrons below 0.01 eV assigned to the subrange R_0.

Inelastic electron–impact cross sections for the 0→1 vibrational excitation in OH, calculated using the method of Riahi et al. [22], are shown by a solid curve. This case was chosen as it provides a smoothly varying curve, so that the accuracy of the simulation can be evaluated with minimum contribution from any structure in the cross sections. The superelastic cross sections for the $1 \to 0$ transition, calculated using Equation (14), are shown by the dashed curve. Using these cross sections in Equation (11), rate constants for the $0 \to 1$ excitation ($\varepsilon = 0.443$ eV) are shown for 2500 logarithmically-spaced initial energies (between 0.005 and 20 eV) by horizontal green lines drawn between the initial and final electron energies at the appropriate vertical position. Rate constants for the superelastic deexcitations are plotted similarly in purple. Transitions ending in the lower-energy ranges all start from within a small energy range near the threshold, so, in applying Equation (12) to calculate the averaged rate constants, the steps in energy must be sufficiently small.

Equations (17), (18), (22) and (23) were applied to calculate the rate constants for transitions between pairs of subranges. These are represented in Figure 2 by left-pointing arrows for inelastic collisions and by right-pointing arrows for superelastic collisions, drawn from the centre of the initial subrange to the centre of the final subrange, with the vertical position showing the value of the average rate constant. At low electron energies, the average rate constants are much lower than the individual ones because, particularly near the threshold, only a fraction of the possible transitions starting within the higher energy subrange end in each lower-energy subrange. Thus, these are better described as "effective" rate constants. This is an example of the way in which details of the cross section within an energy range are embedded in the rate constants. At higher energies, where the transition energy is much less than the width of the energy subranges, the effective rate constants are again lower because only transitions that cross a subrange boundary are included. The effective rate constants for collisions that leave the electron in the same subrange are plotted as octagons for the inelastic collisions and squares for the superelastic collisions.

In some cases, the corrected rate constants produced by Equations (22) and (23) can be negative. While this is unphysical, it makes a necessary correction to maintain conservation of energy. The negative values are indicated by filled symbols in Figure 2.

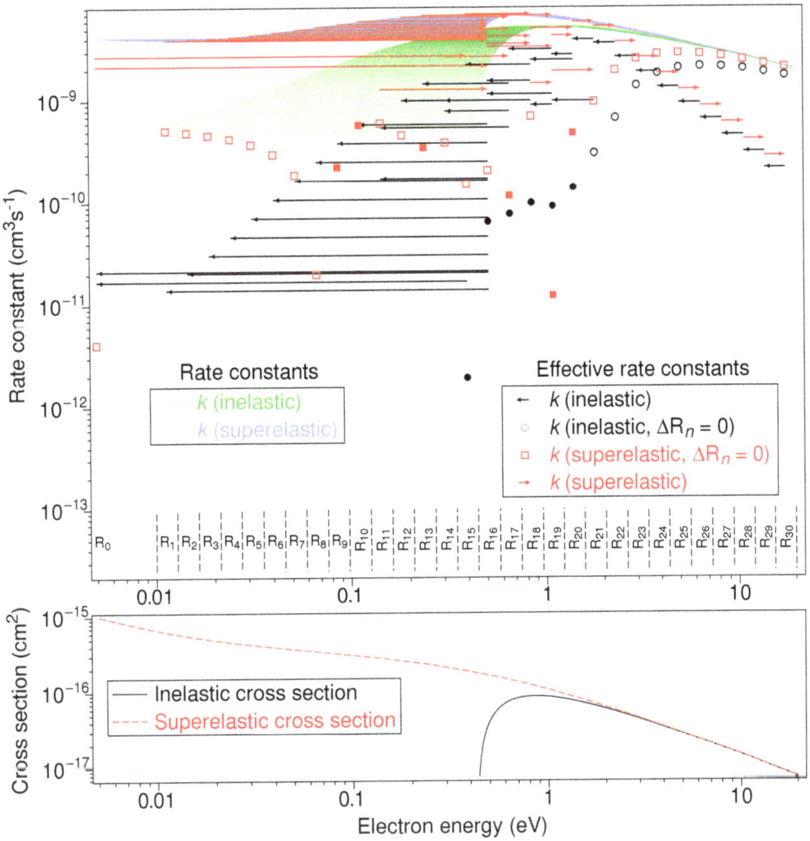

Figure 2. Rate constants (upper panel) and cross sections (lower panel) as a function of electron energy for electron–impact vibrational excitation ($0 \rightarrow 1$) in OH. Rate constants for inelastic collisions are shown by green lines from the initial to the final electron energy. In the simulation (for transitions between pairs of the 31 energy subranges indicated by the vertical dashed lines), the effective rate constants are shown by arrows drawn from the midpoint of the initial subrange to the final, while symbols (circles for inelastic and squares for superelastic collisions) show the effective rate constants for the transitions where the initial and final electron energies are in the same subrange. Negative rate constants are indicated by solid circles for inelastic collisions and solid squares for superelastic collisions.

As examples, the reactions added to the model for inelastic collisions starting in subranges R_{18} and R_{30} are specified as:

$$
\begin{aligned}
\mathrm{OH}(0) + \mathrm{R}_{18} &\xrightarrow{k_{18,14}} \mathrm{OH}(1) + \mathrm{R}_{14} \text{ where } k_{18,14} = 9.921 \times 10^{-10}\ \mathrm{cm}^{-3}\mathrm{s}^{-1} \\
\mathrm{OH}(0) + \mathrm{R}_{18} &\xrightarrow{k_{18,15}} \mathrm{OH}(1) + \mathrm{R}_{15} \text{ where } k_{18,15} = 2.225 \times 10^{-9}\ \mathrm{cm}^{-3}\mathrm{s}^{-1} \\
\mathrm{OH}(0) + \mathrm{R}_{18} &\xrightarrow{k_{18,16}} \mathrm{OH}(1) + \mathrm{R}_{16} \text{ where } k_{18,16} = 1.549 \times 10^{-9}\ \mathrm{cm}^{-3}\mathrm{s}^{-1} \\
\mathrm{OH}(0) + \mathrm{R}_{18} &\xrightarrow{k_{18,18}} \mathrm{OH}(1) + \mathrm{R}_{18} \text{ where } k_{18,18} = 9.862 \times 10^{-11}\ \mathrm{cm}^{-3}\mathrm{s}^{-1} \\
\mathrm{OH}(0) + \mathrm{R}_{30} &\xrightarrow{k_{30,29}} \mathrm{OH}(1) + \mathrm{R}_{29} \text{ where } k_{30,29} = 2.161 \times 10^{-10}\ \mathrm{cm}^{-3}\mathrm{s}^{-1} \\
\mathrm{OH}(0) + \mathrm{R}_{30} &\xrightarrow{k_{30,30}} \mathrm{OH}(1) + \mathrm{R}_{30} \text{ where } k_{30,30} = 1.725 \times 10^{-9}\ \mathrm{cm}^{-3}\mathrm{s}^{-1}
\end{aligned} \tag{24}
$$

A proof-of-concept test of this proposed method, to investigate validity and accuracy, is to simulate the injection of electrons of one energy into a gas of OH molecules that are initially all in the lowest level of the ground state and to consider only excitation to the first vibrational level. While this is unphysical, as it does not consider the electron–electron interactions that would dominate in this case, the rationale is to make a stringent test in a case where a theoretical equilibrium can be calculated. Initially, the electrons will lose energy in exciting the OH molecules, then regain some of it via superelastic collisions, until an equilibrium is reached. The total energy of the system (of electrons plus excited OH molecules) should be equal to the total input energy, while the electrons should have a Maxwell–Boltzmann distribution as given in Equation (13).

Consider that N_e electrons with a total energy E_{tot} are mixed with N_g OH molecules in the lowest level of the ground state. Assuming that at equilibrium the vibrational temperature of the gas and the electron temperature are the same, this temperature T can be found by solution of the equation:

$$\frac{e^{-\varepsilon/(kT)}}{1+e^{-\varepsilon/(kT)}}\varepsilon N_g + 1.5kTN_e = E_{tot}. \tag{25}$$

In order to compare results for logarithmic and linear spacing of the subranges, where the average energy of the electrons E_N (i.e., centre of subrange R_N) is different between the two, N_e is chosen so that the equilibrium gas energy E_g is the same:

$$N_e = \frac{E_g}{E_N - 1.5kT} \text{ where } T = \frac{-\varepsilon}{k\ln\left(\frac{g_f}{1-g_f}\right)},\; g_f = \frac{E_g}{N_g\varepsilon}. \tag{26}$$

The electrons can also gain or lose energy in elastic collisions with the gas molecules. To calculate rate constants for elastic collisions, Equation (15) for electron scattering from O is used, making the further approximation of applying the characteristic temperature T_e of a Maxwellian distribution of electrons to that of a single electron. This gives the change of energy of an electron with energy E (eV) for unit gas and electron densities as:

$$\frac{\Delta E}{\Delta t} = -3.74\times10^{-18}T_e^{1/2}(T_e - T), \tag{27}$$

where $T_e = E/k$. As ΔE is a small fraction of the electron subranges, the rate for transfer from subrange R_j to R_k is divided by the number of individual collisions required to transfer a total energy of $E_j - E_k$, where E_m is the midpoint of the energy subrange R_m. Thus, the rate constant for elastic collisions that transfer energy from range R_j to range R_k is:

$$k_{jk} = \frac{\Delta E}{E_j - E_k} \text{ where} |j-k| = 1. \tag{28}$$

To verify this method, the simulated energy of the electrons can be compared with that predicted in an iteration of the total electron energy E_e:

$$E_e(t+\Delta t) = E_e(t) - \Delta E, \tag{29}$$

where ΔE is calculated using Equation (15).

3. Results

The 119 effective-rate equations illustrated by the arrows and symbols in Figure 2 were applied to a gas containing (per cm^3) 10^8 OH molecules in the ground state and 562,666 electrons in range R_{30} (determined using Equation (26) to given a final OH energy at equilibrium of 9.7 MeV) for a logarithmic distribution of subranges. The simulation was run (with $f = 10^{-6}$) for 1000 s, giving the results shown in Figure 3 for the OH, electron and total energies as a function of time. The simulated values at every thousandth time step are shown by symbols. In addition, the electron distributions at the beginning and the end of the simulation are shown. The theoretical and simulated equilibrium values are plotted as horizontal lines over each energy interval. It can be seen in Figure 3 that energy is transferred from the electrons to the gas molecules, reaching close to an equilibrium after 700 s. At this stage, both the gas energy and the total energy are lower by about 2%, while the electron energy is close to the expected value. The simulated electron energy distribution is close to the calculated Maxwellian distribution ("Equil. ED") but with a residual high-energy tail.

The results for repeating the simulation with 506,677 electrons and a linear distribution of 31 subranges are shown in Figure 4. The simulated OH energy reaches about ∼1% of the equilibrium value after about 400 s, but the electron energy stabilises at (proportionally) a significantly higher absolute value, resulting in a slight excess in the total energy. The excess electron energy appears as higher values in the number of electrons at <1 eV in the electron distribution.

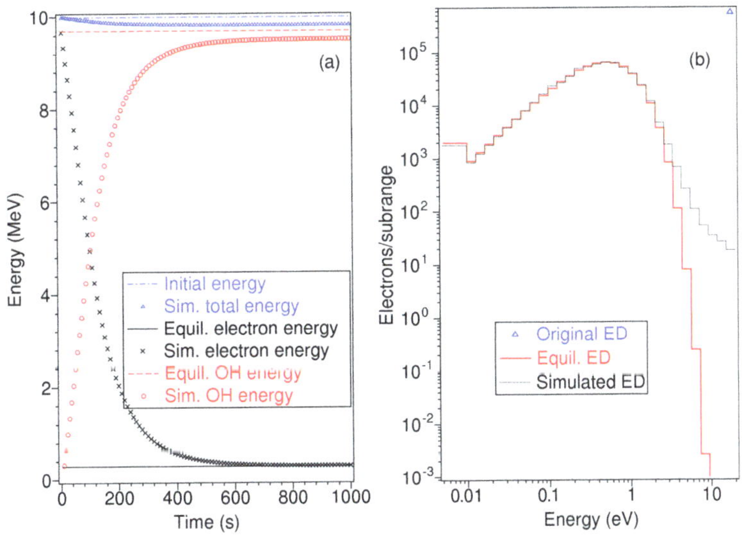

Figure 3. Simulated (symbols) and equilibrium (lines) values of (**a**) gas, electron and total energies for $N = 30$ with a logarithmic distribution of subrange boundaries. The simulated values are plotted as a function of time at every thousandth time step. The electron distributions (**b**) are shown for the original (△), predicted equilibrium (thick line) and simulated at 1000 s (thin line).

In Figure 5, the simulated OH energies are plotted as a function of time for linear and logarithmic spacing of 31 subranges, each for the uncorrected rates (Equations (17) and (18)) and for the rates corrected for energy conservation (Equations (22) and (23)). While there is little difference until about 150 s into the simulation, the values in the corrected cases approach the predicted equilibrium while those for the uncorrected cases rapidly gain excess energy. The energy in the corrected linear case is constant between 500 s and 1,000,000 s, while in the logarithmic case it is constant from 700 s to 500,000 s and then declines slightly. In addition, the OH energies for the corrected logarithmic case with the negative rates set to zero are shown. These show excess energy appearing after about 1000 s.

The simulations summarised in Figure 5 were repeated for $N = 100$, giving the energies plotted in Figure 6 as a function of time. In both the corrected linear and logarithmic cases, the OH energies converge to the predicted value and the equilibrium is then maintained for 1,000,000 s, while for the uncorrected cases the energies still diverge, or go to zero in the case of setting negative rates to zero. The rise time for the logarithmic case is slightly longer than for the linear case.

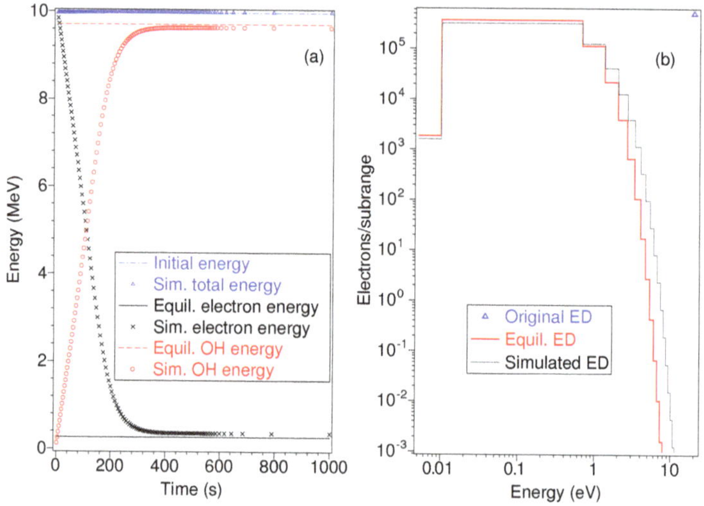

Figure 4. Simulated (symbols) and equilibrium (lines) values of (**a**) gas, electron and total energies for $N = 30$ with a linear distribution of subrange boundaries The simulated values are plotted as a function of time at every thousandth time step. The electron distributions (**b**) are shown for the original (△), predicted equilibrium (thick line) and simulated at 1000 s (thin line).

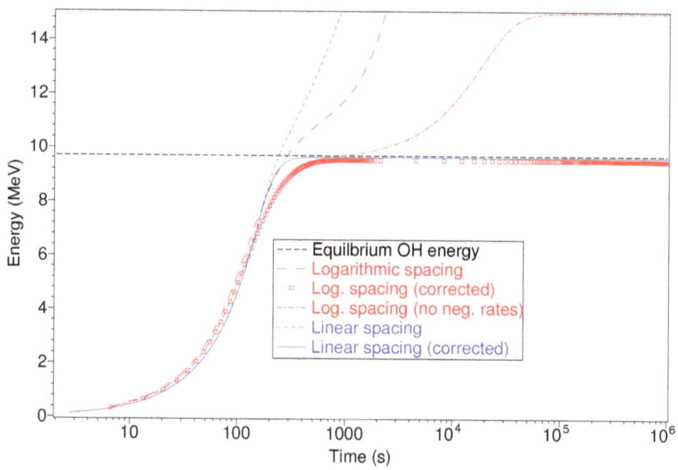

Figure 5. The calculated OH energies as a function of time are shown for logarithmic and linear distributions of 31 subranges, compared with the equilibrium value (– – –). For the linear distribution, uncorrected (- - - -) and corrected (—) cases are shown, while, for the logarithmic distribution, uncorrected (— —), corrected (□) and corrected with negative rates reset to zero (–·–·–) are shown.

In Figure 7, the OH, electron and total energies are plotted as a function of time, for $N = 100$ with logarithmic spacing, showing all approaches and then remaining very close to the predicted values. The percentage errors in the total energy are plotted, showing a rise to about 0.09% before falling to 0.03% when equilibrium is attained. (The discontinuity at about 200 s is due to a change in sign in the difference between the simulated and calculated values).

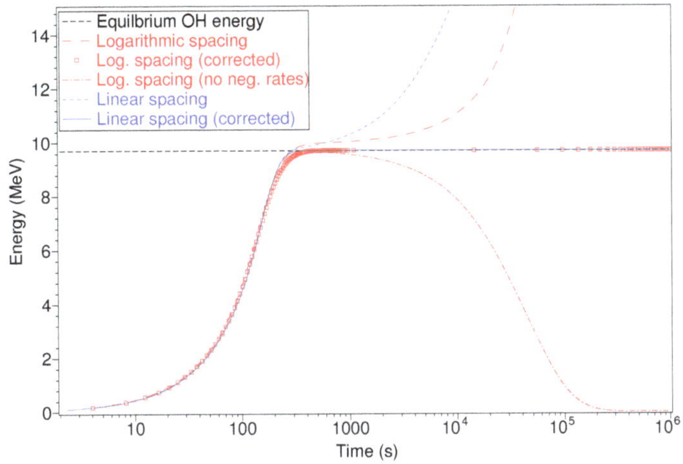

Figure 6. The calculated OH energies as a function of time are shown for logarithmic and linear distributions of 101 subranges, compared with the equilibrium value (– – –). For the linear distribution, uncorrected (- - - -) and corrected (—) cases are shown, while for the logarithmic distribution uncorrected (— —), corrected (□) and corrected with negative rates reset to zero (–·–·–) are shown.

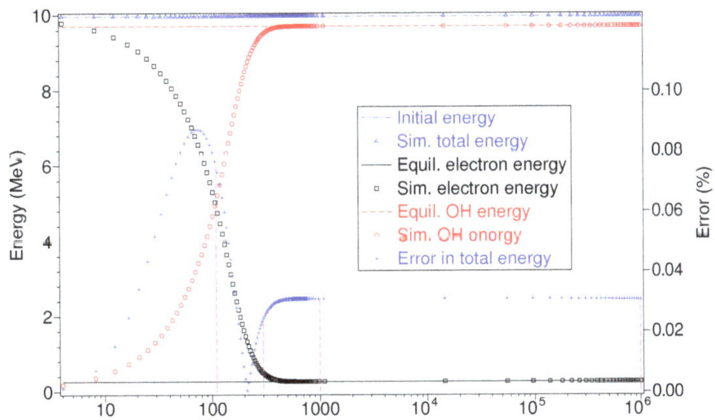

Figure 7. Initial or equilibrium (horizontal lines) and simulated (symbols) energies plotted as a function of time for $N = 100$. The percentage errors in the total energy are shown by plus signs. Vertical dashed lines indicate the times for which the electron distributions are plotted in Figure 8.

The simulated electron distributions, at the times shown by the vertical dashed lines in Figure 7, are plotted along with the theoretical equilibrium distribution in Figure 8. The errors (being the difference between the simulated and theoretical distribution as a percentage of the higher value) are plotted as crosses where the simulated values are higher and circles for where they are lower. At 110 s, most of the electrons are still at high energy, but, by 300 s, the Maxwellian distribution has appeared, with a remaining high-energy tail. This tail, which is almost gone by 1000 s, has disappeared entirely by 10^6 s, at which time the rest of the simulated distribution is unchanged since 1000 s, seen clearly in the similarity of the details of small-scale discrepancies. A possible reason for this remaining small-scale structure is an aliasing effect between the fixed value of ε and the changing sizes of the subranges that result in varying proportions of electrons transferring to a lower or higher-energy subrange.

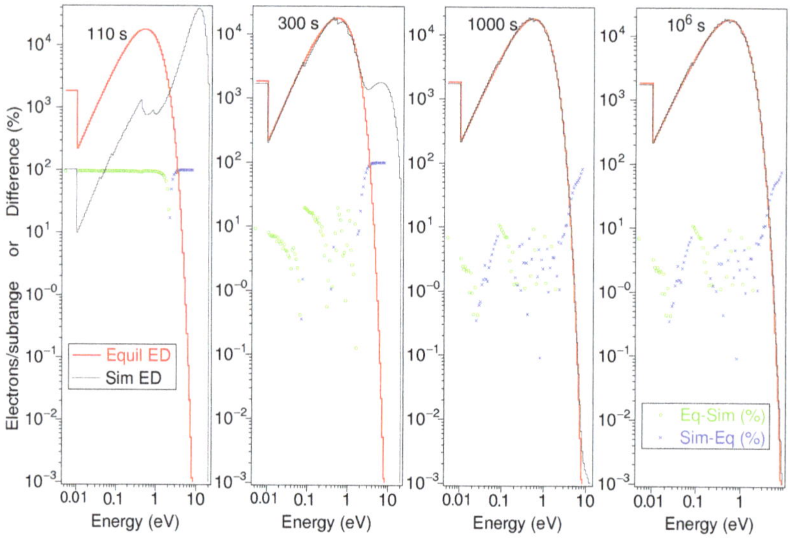

Figure 8. Simulated electron distributions (at 110 s, 300 s, 1000 s and 10^6 s) are plotted along with the predicted equilibrium distribution (thick lines). Errors, being the difference between simulated and theoretical values as a percentage of the larger value, are shown by crosses where the simulated value is larger and circles where it is smaller.

The simulation was applied for different values of the maximum subrange number N and the scaling fraction f. The results are summarised in Table 1, which gives the consequent computational parameters (the number of rate equations N_{eqn}, the number of timesteps in the simulation N_{ts} and, as an indicator of the computational load, their product) and the errors at 10^6 s in the electron energy, OH energy and total energy. The percentage errors in the total energy are plotted in Figure 9 as a function of the computational load ($N_{eqn}N_{ts}$).

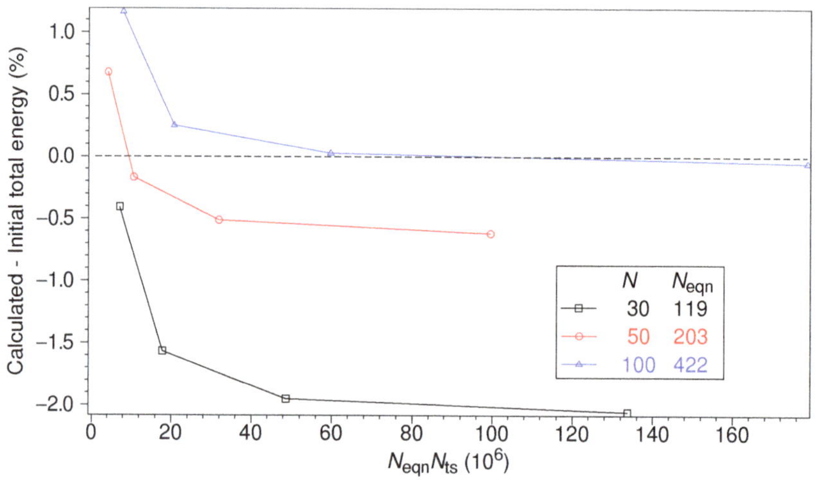

Figure 9. Percentage errors in the total energy for N = 30, 50 and 100, for in each case (left to right) $f = 10^{-4}$, 10^{-5}, 10^{-6} and 10^{-7}, plotted as a function of the computational load ($N_{eqn}N_{ts}$).

Table 1. Summary of consequent computational parameters (number of rate equations N_{eqn}, number of time steps N_{ts} and their product) and percentage errors in the calculated values (of electron energy, OH energy and total energy) resulting from combinations of the initial computational parameters (maximum subrange N and scaling fraction f).

Computational Parameters							
Initial		Consequent			Percentage Errors in Energies:		
N	f	N_{eqn}	N_{ts}	$N_{eqn}N_{ts}$ (10^6)	Electron	OH	Total
30	0.0001000	119	60,489	7.2	5.48	−0.59	−0.41
30	0.0000100	119	149,550	17.8	4.34	−1.75	−1.57
30	0.0000010	119	408,678	48.6	3.93	−2.13	−1.95
30	0.0000001	119	1,123,618	133.7	3.80	−2.23	−2.05
50	0.0001000	203	19,753	4.0	2.41	0.63	0.68
50	0.0000100	203	51,978	10.6	1.73	−0.22	−0.17
50	0.0000010	203	156,367	31.7	1.45	−0.57	−0.51
50	0.0000001	203	489,639	99.4	1.36	−0.67	−0.61
100	0.0001000	422	18,663	7.9	1.56	1.15	1.17
100	0.0000100	422	48,555	20.5	0.84	0.24	0.25
100	0.0000010	422	141,089	59.5	0.67	0.01	0.03
100	0.0000001	422	423,030	178.5	0.60	−0.07	−0.05

In Figure 10, rate equations based on Equation (28) are added to the model to include elastic scattering. The peak OH energy produced by excitation is reduced by about 10%. The OH energy then declines as it is transferred to the electrons and then to the thermal energy of the gas by elastic scattering. As a check on the approximation embodied in Equation (28), Equation (15) is applied at each time step to calculate the energy loss using the instantaneous electron energy, assuming a Maxwellian distribution. The total energy from the simulation ("Sim. total energy") and that from assuming a standard Maxwellian distribution ("Total energy—Maxwellian") are close, particularly in the first 100 s, indicating that the approximations involved in Equation (28) are valid.

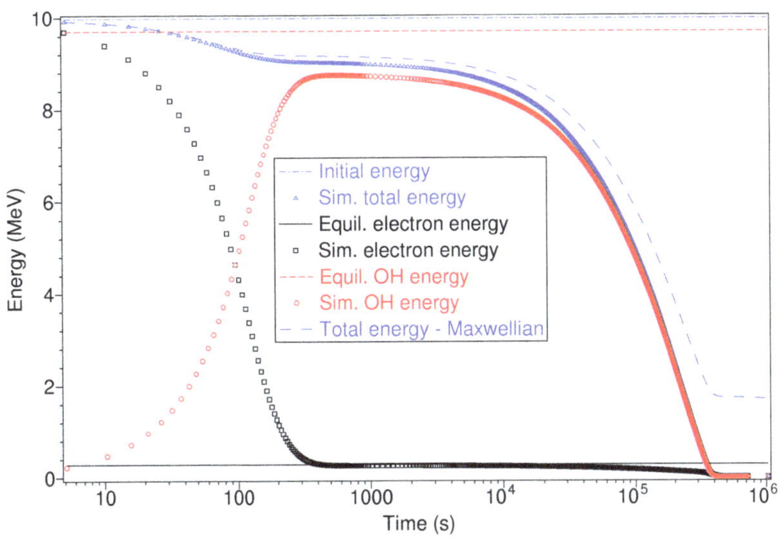

Figure 10. The previous simulation with elastic scattering from the gas molecules added. The total energy (vibrational + electron energy) is shown for: (△) implementation of Equation (28) as rate equations in the simulation and (— —) point-by-point application of Equation (15).

4. Discussion

In Figure 6, it is shown that, with the energy range split into a sufficiently large number of subranges, the simulated OH energy approaches close to the predicted value and is maintained at that value over a long time. In Figure 7, it is shown that the OH, electron and total energies all converge to the predicted values, while Figure 8 shows the simulated electron distribution to be close to the theoretical distribution. Thus, Equations (22) and (23) and their implementation are confirmed to be valid. This is emphasised in Figure 6 by the gross departures from equilibrium in cases where the corrections for conservation of energy are not applied.

Comparison of Figures 5 and 6 shows that, while the rise time is longer for both the linear and logarithmic corrected cases for a smaller number of subranges, the effect is more pronounced in the logarithmic case. As this lag in rise time is not seen for the uncorrected logarithmic case with $N = 30$, where it is close to the values of the linear case until the predicted equilibrium value is reached, it can be inferred that the lag is associated with the correction for energy conservation and that it is more pronounced in the logarithmic case.

As a major rationale for this method is to increase computational efficiency, it is essential to reduce the number of subranges to a minimal value. This introduces inaccuracies as shown in Figures 3 and 4. In the linear case, the electron energy is proportionally high, presumably due to the major part of the electron distribution being averaged into one subrange. In the logarithmic case, the OH and total energy are both slightly reduced and the rise time is substantially increased. Thus, a choice must be made between using a logarithmic or a linear spacing, depending on whether rise time or accuracy of the electron spectrum is more important. An approach to this may be to run both a linear and logarithmic model, so that the difference in results gives an indication of the magnitude of the errors.

The near-accurate calculation of the OH energies in Figure 4 demonstrates the successful incorporation of the detail of the electron–impact cross sections into a few rate equations for transitions in and out of the one subrange [0.01–0.8] eV, which incorporates almost all the electron–impact cross sections below the peak value. This demonstrates the potential of this method to reduce computation times, relative to recalculating the excitation rates for each iteration of the electron spectrum.

In Figure 9, it is seen that, for $N = 100$ and $f < 10^{-6}$, very high accuracy in the equilibrium total energy is maintained to 10^6 s. For each value of N, the discrepancy asymptotes to a fixed value with decreasing f, showing that the time-step method is not a source of error in the results for $f < 10^{-6}$. The figure also shows that there is a substantial increase in accuracy in reducing f to 10^{-6}, but a further decrease makes little difference to the accuracy while substantially increasing the computational load. Thus, to achieve a desired level of accuracy for this particular application, the strategy would be to use $f = 10^{-6}$ and then choose the appropriate number of subranges. The inaccuracy introduced by reducing N is substantially greater than by increasing f. Thus, it appears that an optimum value of f, in a compromise between computation time and accuracy, is 10^{-5}.

Figure 10 shows that elastic scattering (calculated using values for O atoms) reduces the initial excitation of OH by about 10%. Thus, elastic scattering for all the species present must be included in a simulation.

5. Conclusions

A method to simulate nonequilibrium interactions of electrons with gas molecules was proposed and tested. In this method, the energy range of the electrons is split into subranges that are then treated in a time-step calculation in the same way as chemical species, so the electron interactions can be incorporated easily into existing simulations without new coding being required. It was found that, in excitation of gas molecules with one vibrationally-excited level, the initial energy of the electrons was transferred to the gas molecules until an equilibrium was reached that, with sufficiently small subranges, was very close to the predicted equilibrium values. This equilibrium was then maintained over a long time (10^6 s), validating the method of calculating the rates for the electron interactions. The simulated electron spectrum was also very close to the predicted Maxwellian distribution. On reducing the number of energy subranges (which is desirable for reduced computation time), two discrepancies became apparent. For a linear spacing of the subranges, the simulated electron distribution had a higher energy, presumably because most of the equilibrium electron distribution was represented by a single subrange. However, the accuracy of the results despite this low resolution confirmed the potential of the method to reduce computation times by incorporating details of the cross-section spectrum into a small number of rate equations. For a logarithmic spacing, the rise in gas energy was slow, and the final gas and total energies were slightly less than the predicted values. Thus, the proposed method is capable of producing accurate results, but the minimum number of subranges, and thus computational efficiency, will need to be assessed for the requirements and circumstances of the particular application.

Author Contributions: Conceptualization, L.C., D.L.M. and M.J.B.; methodology, L.C. and M.J.B.; software, L.C.; validation, L.C., M.J.B. and D.L.M.; formal analysis, L.C.; investigation, L.C.; resources, M.J.B.; data curation, L.C.; writing—original draft preparation, L.C.; writing—review and editing, M.J.B. and D.L.M.; visualization, L.C.; supervision, M.J.B.; project administration, M.J.B.; funding acquisition, M.J.B. All authors have read and agreed to the published version of the manuscript.

Funding: This work was financially supported by the Australian Research Council (Projects No. DP190100696 and No. DP180101655).

Data Availability Statement: Not applicable.

Conflicts of Interest: The authors declare no conflict of interest.

References

1. Cleary, D.D. Daytime high-latitude rocket observations of the NO, γ, δ and ϵ bands. *J. Geophys. Res.* **1986**, *91*, 11337–11344.
2. Moses, J.I.; Fouchet, T.; Bézard, B.; Gladstone, G.R.; Lellouch, E.; Feuchtgruber, H. Photochemistry and diffusion in Jupiter's stratosphere: Constraints from ISO observations and comparisons with other giant planets. *J. Geophys. Res.* **2005**, *110*, E08001. [CrossRef]
3. Campbell, L.; Brunger, M.J. Calculated meteoroid production of hydroxyl in the atmosphere of Jupiter. *Icarus* **2019**, *326*, 162–169. [CrossRef]
4. Kim, Y.H.; Fox, J.L. The chemistry of hydrocarbon ions in the Jovian atmosphere. *Icarus* **1994**, *112*, 310–325. [CrossRef]
5. Campbell, L.; Brunger, M.J. Modelling of plasma processes in cometary and planetary atmospheres. *Plasma Sources Sci. Technol.* **2013**, *22*, 013002. [CrossRef]
6. Campbell, L.; Brunger, M.J. Electron collisions in atmospheres. *Int. Rev. Phys. Chem.* **2016**, *35*, 297–351. [CrossRef]
7. Thomas, L.; Bowman, M.R. Model studies of the *D*-region negative-ion composition during day-time and night-time. *J. Atmos. Terr. Phys.* **1985**, *47*, 547. [CrossRef]
8. Bates, D.R.; Nicolet, M. The photochemistry of atmospheric water vapor. *J. Geophys. Res.* **1950**, *55*, 301. [CrossRef]
9. Muccignat, D.L.; Stokes, P.W.; Cocks, D.G.; Gascooke, J.; Jones, D.B.; Brunger, M.J.; White, R.D. Simulating the feasibility of using liquid micro-jets for determining electron-liquid scattering cross-sections. *Int. J. Mol. Sci.* **2022**, *23*, 3354. [CrossRef] [PubMed]
10. Muccignat, D. Monte Carlo Simulation of Non-Equilibrium Electron-Liquid Transport. [Undergraduate Honours Thesis]. 2018. Available online: https://www.researchgate.net/publication/360333604_Monte_Carlo_simulation_of_non-equilibrium_electron-liquid_transport (accessed on 7 June 2022).
11. Boyle, G.J.; Tattersall, W.J.; Cocks, D.G.; Dujko, S.; White, R.D. Kinetic theory of positron-impact ionization in gases. *Phys. Rev. A* **2015**, *91*, 052710. [CrossRef]
12. Boudaïffa, B.; Cloutier, P.; Hunting, D.; Huels, M.A.; Sanche, L. Resonant Formation of DNA Strand Breaks by Low-Energy (3 to 20 eV) Electrons. *Science* **2000**, *287*, 1658. [CrossRef] [PubMed]
13. Hahn, M.B.; Dietrich, P.M.; Radnik, J. In situ monitoring of the influence of water on DNA radiation damage by near-ambient pressure X-ray photoelectron spectroscopy. *Commun. Chem.* **2021**, *4*, 50. [CrossRef]
14. Capitelli, M.; Armenise, I.; Bruno, D.; Cacciatore, M.; Celiberto, R.; Colonna, G.; Pascale, O.D.; Diomede, P.; Esposito, F.; Gorse, C.; et al. Non-equilibrium plasma kinetics: A state-to-state approach. *Plasma Sources Sci. Technol.* **2007**, *16*, S30. [CrossRef]
15. Cariolle, D.; Moinat, P.; Teyssèdre, H.; Giraud, L.; Josse, B.; Lefèvre, F. ASIS v1.0: An adaptive solver for the simulation of atmospheric chemistry. *Geosci. Model Dev.* **2017**, *10*, 1467–1485. [CrossRef]
16. Verwer, J.G. Gauss-Seidel iteration for still ODES from chemical kinetics. *SIAM J. Sci. Comput.* **1994**, *15*, 1243. [CrossRef]
17. Tipler, P.A.; Mosca, G. *Physics for Scientists and Engineers*; W. H. Freeman and Company: New York, NY, USA, 2008.

18. Newton, G.P.; Walker, J.C.G.; Meijer, P.H.E. Vibrationally excited nitrogen in stable auroral red arcs and its effect on ionospheric recombination. *J. Geophys. Res.* **1974**, *79*, 3807. [CrossRef]
19. Makabe, T.; Petrović, Z. *Plasma Electronics: Applications in Microelectronic Device Fabrication*; Taylor & Francis Group: New York, NY, USA; London, UK, 2006.
20. Chakrabarti, K.; Laporta, V.; Tennyson, J. Calculated cross sections for low energy electron collision with OH. *Plasma Sources Sci. Technol.* **2019**, *28*, 085013. [CrossRef]
21. Banks, P. Collision frequencies and energy transfer. Electrons. *Planet. Space Sci.* **1966**, *14*, 1085. [CrossRef]
22. Riahi, R.; Teulet, P.; Ben Lakhdar, Z.; Gleizes, A. Cross-section and rate coefficient calculation for electron impact excitation, ionisation and dissociation of H_2 and OH molecules. *Eur. Phys. J. D* **2006**, *40*, 223–230. [CrossRef]

atoms

MDPI

Article

TDCS Calculation for the Ionization of Nitrogen Molecule by Electron Impact

Alpana Pandey and Ghanshyam Purohit *

Department of Physics, University College of Science, Mohanlal Sukhadia University, Udaipur 313001, India; alpana0129@gmail.com
* Correspondence: ghanshyam.purohit@mlsu.ac.in

Abstract: Triple differential cross section (TDCS) results are reported for the electron impact ionization of nitrogen molecules. The TDCSs have been calculated in distorted wave Born formalism using orientation averaged molecular orbital (OAMO) approximation. The TDCS results are presented as average and weighted sum for the outer molecular orbital $3\sigma_g$, $1\pi_u$, $2\sigma_u$ and the inner $2\sigma_g$ molecular orbital. The obtained theoretical TDCSs are compared with the available measurements. The results are analysed in terms of the positions and relative intensities of binary and recoil peaks. Within a first order model and for a complex molecule, a reasonable agreement is obtained with the experimental data in the binary peak region with certain discrepancies in position and magnitude in the recoil peak region.

Keywords: TDCS; weighted and average sum; DWBA; PCI; correlation polarization

Citation: Pandey, A.; Purohit, G. TDCS Calculation for the Ionization of Nitrogen Molecule by Electron Impact. *Atoms* **2022**, *10*, 50. https://doi.org/10.3390/atoms10020050

Academic Editor: Jean-Christophe Pain and Himadri Chakraborty

Received: 18 April 2022
Accepted: 14 May 2022
Published: 18 May 2022

1. Introduction

Over the past three decades, electron impact ionization studies have gained considerable experimental and theoretical attention. Atomic and molecular physicists consider electron impact ionization of targets such as atoms, ions and molecules to be one of the most important collision processes. Electron impact ionization also referred to as (e, 2e) [1] involves the collision of a projectile (incident electron) with a target (either an atom or an ion or a molecule) leading to the ionization of the target. Upon determining the energies and the momenta of all the particles involved in the collision, one can have a complete understanding of the ionization process. Thus, (e, 2e) collisions have become an important tool for investigating the collision dynamics of targets. The triple differential cross section (TDCS) is the physical quantity that is of prime interest in these studies, it provides information about collision processes, ionization mechanisms, and the dependence of the ionization process on the electron kinematics under which ionization is taking place. Experimental techniques such as cold target recoil-ion momentum spectroscopy (COLTRIM) [2] and recoil-ion spectroscopy (RMS) [3] have been instrumental to obtain the cross-sections of the electron impact ionization of atomic and molecular targets.

Most of the experimental efforts have been focused on inert gas atoms [4], and limited attempts have been made to measure cross-sections of molecular targets and the theoretical support also has been less for these measurements. It is mainly due to the experimental difficulties in measuring TDCS for molecules, as a result of the close spacing of the different molecular states, and the difficulty of developing a theoretical model to explain molecular ionization, as a consequence of the multi-centre nature of the target. There have been efforts to study the electron impact ionization of molecular targets [5–8]. The TDCS has been calculated for various molecular targets, from simple diatomic atoms to very complex molecules. Some of these studies may be mentioned; H_2 [9–11], N_2 [12–17], CO_2 [18], H_2O [19–21] and a wide range of biologically complex molecules like pyrimidine [22], thymine [23] etc. Electron impact single ionization cross-sections of molecular targets can be calculated by using various theoretical techniques. One of the most successful theoretical models to study electron impact ionization of various targets is the distorted-wave Born approximation (DWBA) [24]. The single ionization of a complex target can be viewed as a three-body problem in which the spectator electrons are represented by spherically symmetric potentials. In the DWBA formalism, these spherically symmetric potentials can be utilized in Schrödinger equation to calculate the continuum wave functions. In the Born approximation, it is not possible to obtain exact solutions to the Schrödinger equation. Therefore, it is impossible to describe all the interactions and processes that may take place during ionization. It is possible to modify the theoretical formalism in several ways, such as by approximating the post-collision interaction (PCI), taking into account correlation-polarization effects, and considering the electron exchange.

In the low to intermediate impact energy region, diatomic molecules H_2 and N_2 and the triatomic H_2O and CO_2 are ionized as given in most of the recent studies. Several studies [25–27] have been done for N_2, where TDCS is calculated at different projectile energy. The electron impact cross sections of N_2 molecules have been calculated using the distorted wave Born approximation (DWBA) method [28]. DWBA has been found to give a reasonable agreement with the measurements for the (e, 2e) studies on molecules with certain discrepancies, particularly in the recoil peak region.

In the present communication, we investigate the ionization of nitrogen molecules at different energies within the distorted-wave Born approximation formalism using the orientation-averaged molecular orbital approximation. We report the TDCS results for the ionization of nitrogen molecules at scattered electron energy 500 eV, for the coplanar asymmetric emission of electrons [29]. In the present study, we have used atomic units ($\hbar$ = e = m_e = 1) for all calculations. In the next section, we outline the theoretical approach used to calculate TDCS.

2. Theory

The initial—state Hamiltonian chosen in the standard DWBA is given by

$$H_0 = H_{target} + T_p + U_i \quad (1)$$

where H_{target} is the Hamiltonian for the neutral target, T_p is the kinetic energy operator for the projectile and U_i is an initial-state spherically symmetric potential for the ionization process. The DWBA approach was generalized to molecules [30,31].

The triple differential cross section for the ionization of nitrogen molecule by electron impact is given by

$$\frac{d^3\sigma}{dk_s dk_e dE_e} = (2\pi)^4 \frac{k_s k_e}{k_i} |t|^2, \quad (2)$$

where k_s, k_e, k_i are the momenta of the scattered, ejected and the incident electrons, respectively. The term E_e and 't' is referred to as ejected electron energy and transition matrix element respectively. The transition matrix is represented in the terms of direct and exchange scattering amplitude. The amplitude is given by

$$|t|^2 = |f_{dir}|^2 + |f_{ex}|^2 - |f_{dir}||f_{ex}| \quad (3)$$

where the direct scattering amplitude (f_{dir}) is given by

$$f_{dir} =< X_s(r_1)X_e(r_2)|\frac{(-Z)}{r_{12}}|X_b(r_2)X_i(r_1) > \quad (4)$$

Similarly, the exchange term (f_{ex}) can be expressed as

$$f_{ex} =< X_e(r_1)X_f(r_2)|\frac{(-Z)}{r_{12}}|X_b(r_2)X_i(r_1) > \quad (5)$$

In this equation, r_1, r_2 are the position vectors for projectile and active electron, and r_{12} is the distance between projectile and active target electron. For incident, bound, and scattered electrons, X_i, X_b, X_s represent their distorted wave functions, respectively. The bound state for the orbitals of the N_2 molecule is approximated as the orientation averaged molecular orbital where. The molecular wave function has been calculated using the B3LYP/TZ2P [32] basis sets based on density functional theory. We calculate the distorted waves on symmetric potential, which is based on the Hartree-Fock charge distribution for N_2 averaged over all molecular orientations. For the incoming electron wave function, the Schrödinger equation is given by

$$(T + U_i - \frac{k_i^2}{2})x_i(k_i, r) = 0 \tag{6}$$

T represents the kinetic energy operator. In the initial state, the distortion potential is determined by the nuclear contribution and the electronic contribution i.e.,

$$U_{static} = U_{el} + U_{nuc} \tag{7}$$

By averaging the two N_2 nuclei over all orientations, we get the nuclear part. It is obtained as a result of placing the nuclear charge on a spherical shell with a radius equal to the distance of the nucleus from the centre of mass. From the calculated molecular charge density averaging over all angular orientations, the electronic part is derived. The final state distorted potential is generated in a similar way except that the active electron is removed from the charge distribution. The distorting potential proposed by Furness and McCarthy [33], which was later corrected by Riley and Truhlar [34] is added to the static Hartree—Fock distorting potential. The exchange—distortion potential U_E generated for same is given by

$$U_E = 0.5[E_0 - U_{static}(r) - (E_0 - U_{static}(r))^2 + 4\pi\rho(r)] \tag{8}$$

To calculate the TDCSs, we have also included the correlation polarization potential U_{CP} in the distorting potential which is given by

$$U_{CP} = U_{SR}^{Corr}(r), r \leq r_0 \tag{9}$$

$$= \frac{-\alpha_d}{2r^4}, r > r_0 \tag{10}$$

$U_{SR}^{Corr}(r)$ is the short range correlation potential [35] and α_d dipole polarizability of the target.

We have treated the post collision interaction between the two outgoing particles by the Ward-Macek approximation [36]. In the ward-Mack approximation, one replaces the actual final state e-e separation r_{12} by an average value directed parallel to k_{12}. The average separation is given by

$$r_{12}^{ave} = \frac{\pi^2}{16}(1 + \frac{0.627}{\pi}\sqrt{\epsilon} ln\epsilon)^2 \tag{11}$$

ϵ is the total energy of the two exiting electrons. With this approximation, the ward-Macek factor is given by

$$M_{ee} = \frac{\gamma}{\exp(\gamma) - 1}|_1F_1(i\lambda, 1, -2ik_{12}r_{12}^{ave}) \tag{12}$$

Where, $\frac{\gamma}{\exp(\gamma)-1}$ = N_{ee} ,which is the Gamow factor.

Also, γ = $\frac{-2\pi}{k_1-k_2}$, $\lambda = \frac{1}{k_1-k_2}$

The present model DWBA is employed to calculate the TDCS for the ionization of the outer $3\sigma_g$, $1\pi_u$, $2\sigma_u$ and the 'inner' $2\sigma_g$ molecular orbital of the nitrogen molecule by electron impact. The results and discussion is presented in the next section.

3. Results and Discussion

In the present study, we report the results of TDCS calculated in the distorted wave Born approximation approach for the electron impact single ionization of nitrogen molecules which includes orientation-averaged molecular orbital approximation. We have calculated TDCSs for the ionization of nitrogen molecules from 'outer' valence orbital $3\sigma_g$, $1\pi_u$, $2\sigma_u$ and the 'inner' $2\sigma_g$ molecular orbital, these orbital have 15.6, 16.7, 18.75 and 39.9 eV ionization potentials respectively. As the electronic states of the three outer orbital are very closely spaced, it is difficult to be resolved. Therefore, the weighted sum of TDCS for $3\sigma_g$, $1\pi_u$, $2\sigma_u$ orbital with relative efficiencies of 1, 0.78, and 0.32 have also been plotted.

The angular distribution of TDCSs are reported for scattered electron energy E_s = 500 eV and scattering angle -6^0 at ejected electron energies E_e = 37 eV, E_e = 74 eV and E_e = 205 eV and the obtained results are compared with the available measurements [29]. The experimental TDCSs [29] and TDCSs obtained by present study are analyzed in terms of the magnitude of binary and recoil peak and their respective positions. Along with the peak positions and intensity, the recoil-to-binary peak ratios have been obtained and compared for measurements [29] as well as for the theoretically obtained results. The TDCSs have been computed in standard DWBA formalism with first Born term. TDCS calculations are also reported including the correlation-polarization potential and PCI effects in the standard DWBA.

The present results are plotted in Figures 1–3 for ejected electron energies E_e = 37 eV, E_e = 74 eV and E_e = 205 eV. The solid red circles are the experimental TDCS [29]. The solid curve is for DWBA results with first Born term calculated for the average sum of the orbitals. The dashed curve is DWBA results including correlation polarization potential, and the dotted curve represents DWBA calculations including polarization potential and PCI effects. The dark circles, upside triangles and hollow circles represent the standard DWBA results, DWBA with polarization potential and DWBA with polarization potential and PCI with the weighted sum of the orbital respectively.

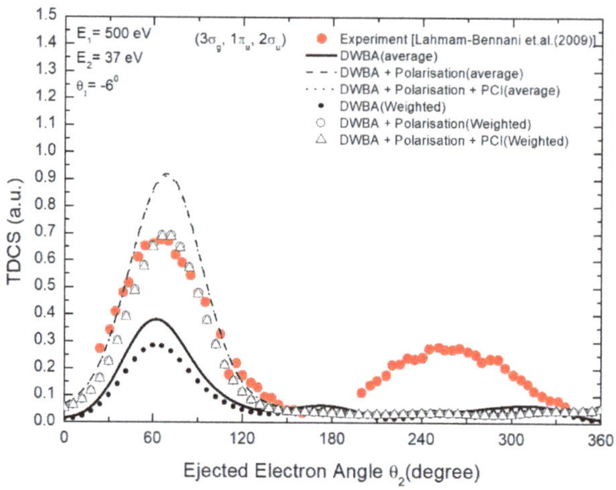

Figure 1. Electron-impact TDCS for N_2 molecule calculated for the weighted and average sum of the outer orbital ($3\sigma_g$, $1\pi_u$, $2\sigma_u$). The ejected electron energy is 37 eV. Kinematics and legends used are displayed in the figure frame.

The TDCS results are presented for ejected electron energy 37 eV in Figure 1. A good agreement with experimental results [29] has been obtained in the present standard DWBA with first Born term as well as the DWBA results including correlation-polarization potential and PCI in the binary peak region, particularly in terms of the binary peak positions. The DWBA results calculated by including polarization potential and PCI with weighted sum of orbital are in very good agreement with the measurements [29] in terms of binary peak positions as well as magnitude of the binary lobe. In the binary peak region, the standard DWBA results for average and weighted sum shows the binary peaks at ejected electron angle 61° and 62° in comparison to the experimental binary peak obtained at 65°. The binary peaks of the TDCS curve calculated using DWBA including polarization and PCI for both average and weighted sum are shifted towards higher ejected electron angles at 69°. It is observed that the experimental TDCS shows recoil peak in the range of 200°–300° with the recoil peak position around 260° . Theoretical TDCS fails to reproduce the recoil peak in the same range and the magnitude of theoretical recoil peak also does not match with the experimental data.

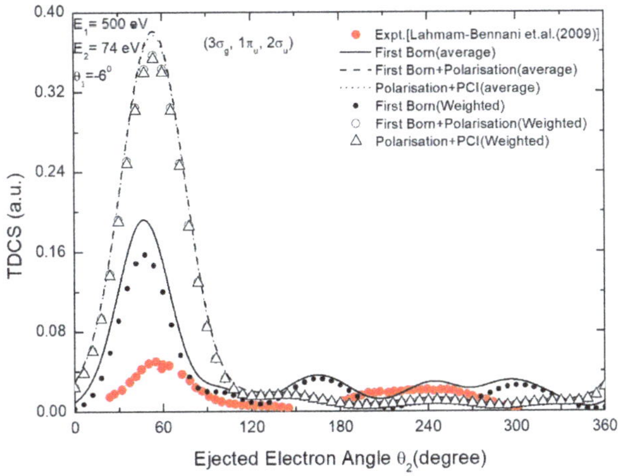

Figure 2. Same as Figure 1. The ejected electron energy is 74 eV.

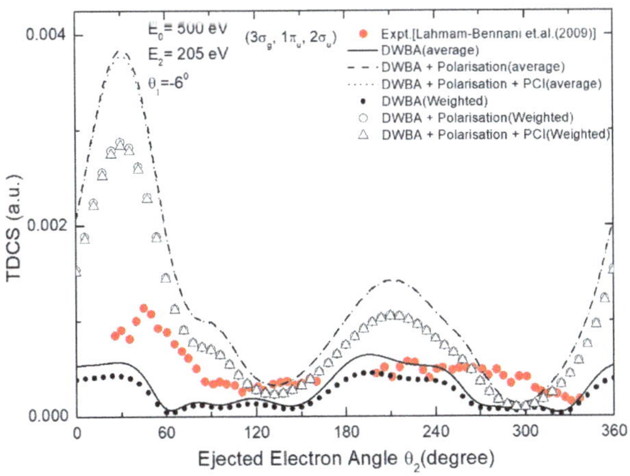

Figure 3. Same as Figure 1. The ejected electron energy is 205 eV.

As shown in Figure 2, the calculated and experimental TDCSs are for an ejected electron energy of 74 eV. Both the weighted and average sums of the binary peak's position show discrepancies in the binary region, and there is less agreement with the experimental data for smaller ejection angles compared with higher ejection angles. It is observed that polarization potential and PCI are not very significant at this ejected electron energy. In terms of binary peak, the theoretical peaks obtained are shifted towards lower ejection angles relative to the experimental binary peak. The recoil to binary peak ratio obtained in the present theoretical TDCS results does not match with the experimental data [29]. The theoretical results with weighted sum including polarization and PCI show a smaller recoil peak, however the standard DWBA results show more than a two-peak structure.

At ejected electron energy E_e = 205 eV, the binary peaks observed in the theoretical TDCSs are shifted towards the lower ejection angle in comparison to the measurements [29]. Both the weighted sum of the orbitals and the average sum of the orbitals fails to reproduce the position of the experimental binary peak and there is also disagreement in the recoil to binary peak ratio. In case of Recoil peak, the theoretical models present better results, however there are discrepancies in the position of the recoil peak. Despite the fact that there are no experimental points for ejection angles smaller than 20°, the measured data show some sign of a binary peak splitting.

Along with weighted and average sum of the outer orbitals, we have analysed TDCS results in terms of $3\sigma_g$, $1\pi_u$, $2\sigma_u$ contribution individually. In Figure 4, the TDCS results are presented for the ejection energy 37 eV. It can be observed that the TDCS corresponding to the $1\pi_u$, including polarization potential is the major contributor to the weighted and average sum of the orbitals in the binary region. There is large discrepancy in the trends of TDCS in the recoil peak region. There is a huge discrepancy in the recoil to binary peak ratio.

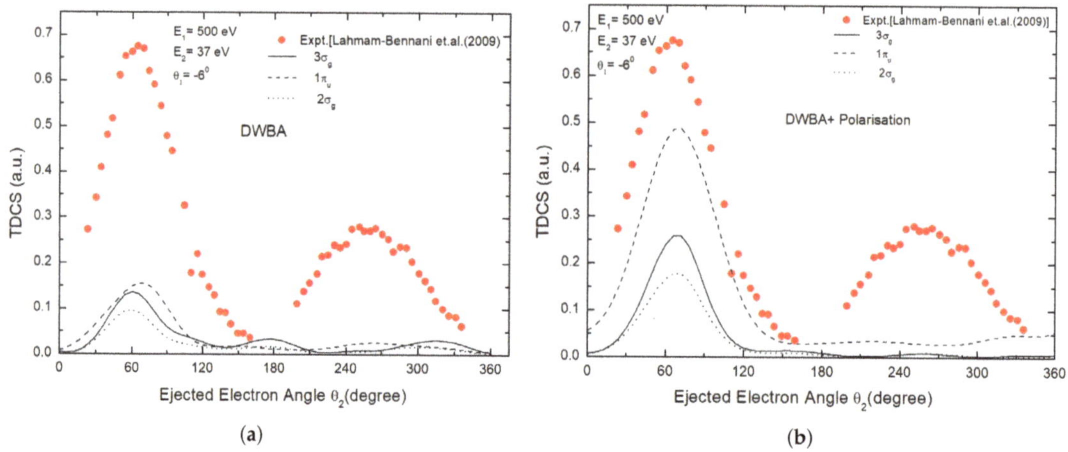

Figure 4. Electron-impact TDCS for N_2 molecule calculated for $3\sigma_g$, $1\pi_u$, $2\sigma_u$ valence orbitals. The ejected electron energy is 37 eV. (**a**) Individual orbital contribution for TDCS calculated by DWBA formalism, (**b**) shows contribution of individual orbital calculated by including polarisation potential. Kinematics and legends used are displayed in the figure frame.

We can see a totally different situation in Figure 5, where we analyse orbitals at ejection energy 74 eV. The size of the contribution of $1\pi_u$, $2\sigma_u$ is nearly equal to the experimental TDCS. Furthermore, it may be observed that the recoil to binary ratio of these orbitals is nearly the same. However, $3\sigma_g$ show different behaviour and fails to reproduce the experimental results.

In Figure 6, the TDCS obtained by DWBA and correlation polarisation potential at ejected energy 205 eV is presented. In both the cases, highest contribution is given by $1\pi_u$, which also gives a higher recoil to binary peak ratio in the case of DWBA, including polarization potential results compared to the $2\sigma_u$, $3\sigma_g$ orbitals, while the recoil peak is small for the DWBA formalism. We can also see a significant difference between these formalisms at the binary peak.

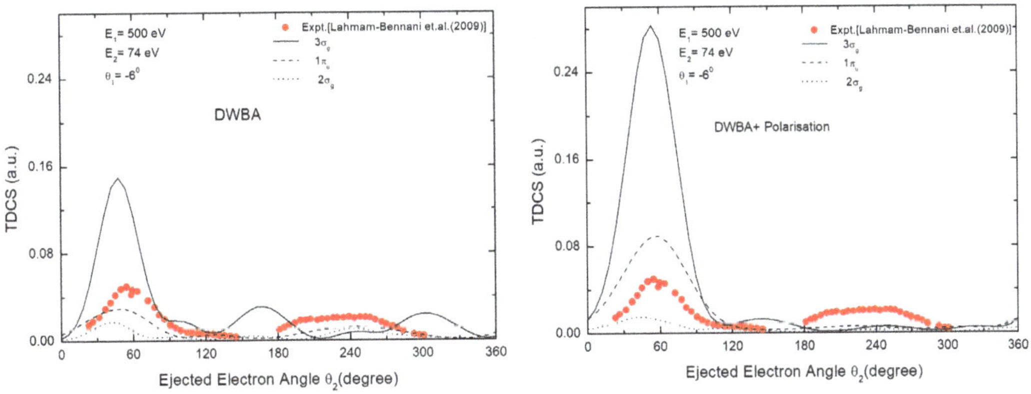

(**a**) (**b**)

Figure 5. Same as Figure 4 at ejected electron energy 74 eV.

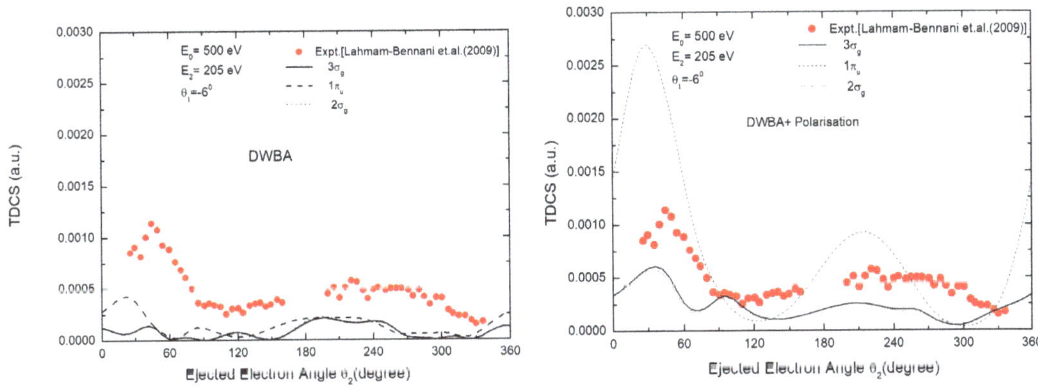

(**a**) (**b**)

Figure 6. Same as Figure 4 at ejected electron energy 205 eV.

Besides, calculating TDCS corresponding to the ionization of nitrogen molecules from the outer orbital, we have also calculated TDCSs for the ionization from inner orbital. TDCSs obtained for the ionization taking place from $2\sigma_g$ orbitals at 36 eV, 74 eV, and 205 eV ejected energies are presented in Figures 7–9. It can be seen that the theoretical models give fair agreement with the experimental data in the binary region. However, significant differences are observed between the theoretical and experimental results in the recoil region. There is no prominent difference between the theoretical TDCS obtained by correlation potential and PCI effect. On the other hand, it appears that experimental binary peak for the inner orbitals in the binary peak region is well reproduced by the TDCS including polarization potential.

TDCSs obtained for 37 eV ejection energy are plotted in Figure 7. The TDCS obtained by including polarization potential shows reasonable agreement with the binary peak of measurements [29] however fails to reproduce the recoil peak as observed in measurements. The theoretical TDCS including polarization potential shows different behaviour for inner and outer orbitals. In case of inner orbitals the TDCS calculated by including polarization potential are in better agreement with the measurements [29].

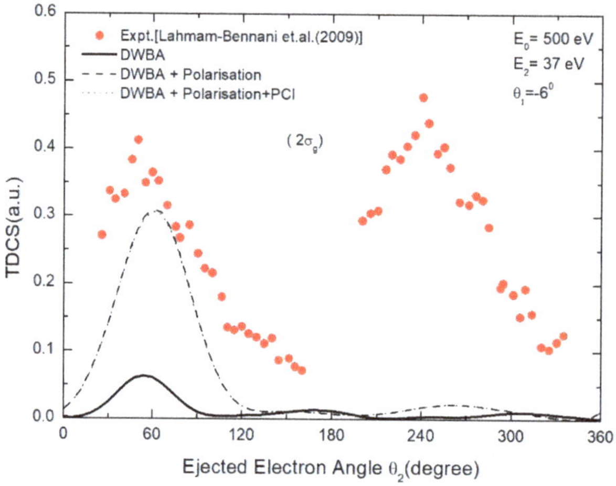

Figure 7. TDCS calculation for the ionization of the inner orbital $2\sigma_g$ of N_2. The ejected electron energy chosen here is 37 eV. The legends and kinematics are displayed in the Figure frame.

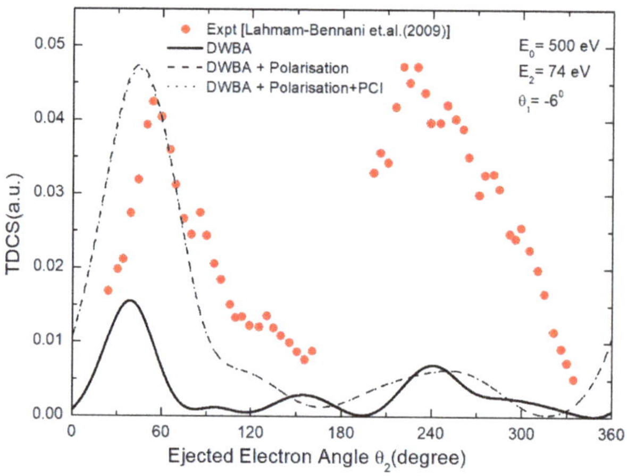

Figure 8. Kinematics is same as Figure 4 and the ejected electron energy chosen is 74 eV.

There is a mismatch between theoretical and experimental data in terms of recoil to binary ratio. All theoretical frameworks fail to reproduce the experimental ratio of recoil to the binary peak.

At high ejection energies, better results are reproduced for the recoil region. At 37 eV, no recoil peak can be seen in theoretical results. For 74 eV a small peak is obtained for recoil region, while at 205 eV a better resemblance between the theoretical and experimental results is obtained in the recoil region. Furthermore, the experimental recoil to binary peak ratio is much greater than the theoretical one for all the kinematics chosen. It is believed that the strong interaction between the ejected electron and the residual ion causes a large recoil intensity. This interaction is enhanced for orbitals with an inner valence and targets with multiple electrons and multiple centres [29]. It is clear that the approximation employed in the present study to calculate TDCS requires more efforts to include effects such as multiple scattering, second order effects to analyse available measurements. We have found the inclusion of target polarization potential to be significant up to an extent; however, the PCI is found to be not very significant for the present kinematics.

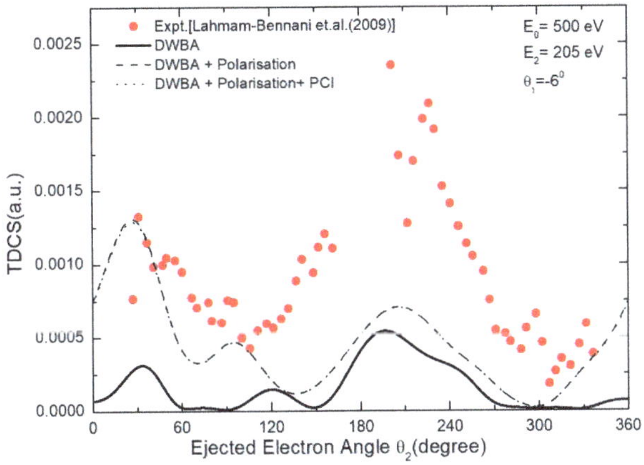

Figure 9. Kinematics is same as Figure 4 and the ejected electron energy chosen is 205 eV.

4. Conclusions

TDCSs have been calculated for electron impact ionization of outer $3\sigma_g$, $1\pi_u$, $2\sigma_u$ and the 'inner' $2\sigma_g$ molecular orbital of nitrogen molecules. The effect of target polarization and post collision interaction between the scattered and ejected electrons after the collision has also been investigated.

The cross sections have been calculated at 37 eV, 74 eV, and 205 eV ejected energies. Due to the close proximity of the outer orbital, the TDCS results are also presented as the average and weighted sum for these orbital. The TDCS results are analysed in terms of positions and relative magnitudes of binary and recoil peaks. For the outer orbital at 37 eV ejected electron energy, polarization potential has been found significant and the results are in good agreement within the binary region however, PCI is not able to make significant changes in the trends of TDCS. At high ejection energy, the theoretical calculations seem to reproduce better results for the recoil region. There is a high discrepancy at recoil peak for 37 eV but gives a qualitative agreement for 74 eV and 205 eV ejection energy.

In the case of ionization from the inner orbital of the nitrogen molecule, large discrepancies are observed between theoretical and experimental results however polarization potential is found to be significant in the binary peak region. There is a qualitative agreement with the binary peak and poor agreement for the recoil peak. The experimental recoil to binary peak ratio is not well produced by the theoretical methods implied.

The current theory-experiment discrepancies seem to motivate applications of more sophisticated methods for improvement. For instance, it would be worthwhile to examine the trends of TDCS of nitrogen molecules further with the distorted wave second Born approximation.

Author Contributions: A.P. and G.P. contributed equally in the writing and analysis of the manuscript. All authors have read and agreed to the published version of the manuscript.

Funding: This research was funded by Science and Engineering Research Board (SERB), New 246 Delhi. Grant No. CRG/2019/001059.

Institutional Review Board Statement: Not applicable.

Informed Consent Statement: Not applicable.

Data Availability Statement: The associated data will be available from authors upon request.

Acknowledgments: We acknowledge the grant received from Science and Engineering Research Board(SERB), New Delhi in the form of SERB CRG project (CRG/2019/001059).

Conflicts of Interest: The authors declare no conflict of interest.

References

1. Lahmam-Bennani, A. Recent developments and new trends in (e, 2e) and (e, 3e) studies. *J. Phys. B At. Mol. Opt. Phys.* **1991**, *24*, 2401. [CrossRef]
2. Dörner, R.; Mergel, V.; Jagutzki, O.; Spielberger, L.; Ullrich, J.; Moshammer, R.; Schmidt-Böcking, H. Cold target recoil ion momentum spectroscopy: A 'momentum microscope'to view atomic collision dynamics. *Phys. Rep.* **2000**, *330*, 95–192. [CrossRef]
3. Ullrich, J.; Moshammer, R.; Dörner, R.; Jagutzki, O.; Mergel, V.; Schmidt-Böcking, H.; Spielberger, L. Recoil-ion momentum spectroscopy. *J. Phys. B At. Mol. Opt. Phys.* **1997**, *30*, 2917. [CrossRef]
4. Naja, A.; Casagrande, E.S.; Lahmam-Bennani, A.; Stevenson, M.; Lohmann, B.; Dal Cappello, C.; Bartschat, K.; Kheifets, A.; Bray, I.; Fursa, D.V. (e, 2e) triple differential cross-sections for ionization beyond helium: The neon case at large energy transfer. *J. Phys. B At. Mol. Opt. Phys.* **2008**, *41*, 085205. [CrossRef]
5. Lahmam-Bennani, A.; Naja, A.; Casagrande, E.S.; Okumus, N.; Dal Cappello, C.; Charpentier, I.; Houamer, S. Dynamics of electron impact ionization of the outer and inner valence (1t2 and 2a1) molecular orbitals of CH_4 at intermediate and large ion recoil momentum. *J. Phys. B At. Mol. Opt. Phys.* **2009**, *42*, 165201. [CrossRef]
6. Mouawad, L.; Hervieux, P.A.; Dal Cappello, C.; El Bitar, Z. Ionization of phenol by single electron impact: Triple differential cross sections. *J. Phys. B At. Mol. Opt. Phys.* **2019**, *53*, 025202. [CrossRef]
7. Singh, P.; Purohit, G.; Champion, C.; Patidar, V. Electron-and positron-induced ionization of water molecules: Theory versus experiment at the triply differential scale. *Phys. Rev. A* **2014**, *89*, 032714. [CrossRef]
8. Wang, Y.; Wang, Z.; Gong, M.; Xu, C.; Chen, X. Theoretical study of (e, 2e) triple differential cross sections of pyrimidine and tetrahydrofurfuryl alcohol molecules using multi-center distorted-wave method. *Chin. Phys. B* **2022**, *31*, 010202. [CrossRef]
9. Al-Hagan, O.; Kaiser, C.; Madison, D.; Murray, A.J. Atomic and molecular signatures for charged-particle ionization. *Nat. Phys.* **2009**, *5*, 59–63. [CrossRef]
10. Pindzola, M.S.; Colgan, J.P.; McLaughlin, B.M. Electron-impact double ionization of the H_2 molecule. *J. Phys. B At. Mol. Opt. Phys.* **2018**, *51*, 035206. [CrossRef]
11. Senftleben, A.; Pflüger, T.; Ren, X.; Al-Hagan, O.; Najjari, B.; Madison, D.; Dorn, A.; Ullrich, J. Search for interference effects in electron impact ionization of aligned hydrogen molecules. *J. Phys. B At. Mol. Opt. Phys.* **2010**, *43*, 081002. [CrossRef]
12. Hargreaves, L.R.; Colyer, C.; Stevenson, M.A.; Lohmann, B.; Al-Hagan, O.; Madison, D.H.; Ning, C. (e, 2 e) study of two-center interference effects in the ionization of N_2. *Phys. Rev. A* **2009**, *80*, 062704. [CrossRef]
13. Toth, I.; Nagy, L. Ionization of molecular nitrogen by electron impact in (e, 2e) processes. *J. Phys. B At. Mol. Opt. Phys.* **2011**, *44*, 195205. [CrossRef]
14. Murray, A.J.; Hussey, M.J.; Gao, J.; Madison, D. States of N_2—comparison between experiment and theoretical predictions of the effects of exchange, polarization and interference. *J. Phys. B At. Mol. Opt. Phys* **2006**, *39*, 3945–3956. [CrossRef]
15. Chaluvadi, H.; Ozer, Z.N.; Dogan, M.; Ning, C.; Colgan, J.; Madison, D. Observation of two-center interference effects for electron impact ionization of N_2. *J. Phys. B At. Mol. Opt. Phys.* **2015**, *48*, 155203. [CrossRef]
16. Tachino, CA and Martín, F and Rivarola, RD Theoretical study of interference effects in single electron ionization of N2 molecules by proton impact. *J. Phys. B At. Mol. Opt. Phys.* **2011**, *45*, 025201. [CrossRef]
17. Ozer, Z.N. Differential cross sections of nitrogen containing molecules at intermediate electron impact energy. *AIP Conf. Proc.* **2018**, *2042*, 020027.
18. Morillo-Candas, A.; Silva, T.; Klarenaar, B.; Grofulović, M.; Guerra, V.; Guaitella, O. Electron impact dissociation of CO_2. *Plasma Sources Sci. Technol.* **2020**, *29*, 01LT01. [CrossRef]
19. Zhou, J.; Ali, E.; Gong, M.; Jia, S.; Li, Y.; Wang, Y.; Zhang, Z.; Xue, X.; Fursa, D.V.; Bray, I.; et al. Absolute triple differential cross sections for low-energy electron impact ionization of biochemically relevant systems: Water, tetrahydrofuran, and hydrated tetrahydrofuran. *Phys. Rev. A* **2021**, *104*, 012817. [CrossRef]
20. Dhankhar, N.; Choubisa, R. Triple differential cross-section for the twisted electron impact ionization of water molecule. *arXiv* **2021**, arXiv:2108.00692.
21. Fernández-Menchero, L.; Otranto, S. Fully and double differential cross sections for the single ionization of H_2O by bare ion impact. *J. Phys. B At. Mol. Opt. Phys.* **2014**, *47*, 035205. [CrossRef]
22. Mouawad, L.; Hervieux, P.A.; Dal Cappello, C.; Pansane, J.; Robert, V.; El Bitar, Z. Triple differential cross sections for the ionization of pyrimidine by electron impact. *Eur. Phys. J. D* **2019**, *73*, 1–8. [CrossRef]
23. Singh, P.; Champion, C. Theoretical study of (e, 2e) triple-differential cross sections for DNA component ionization by electrons and positrons. *J. Phys. B At. Mol. Opt. Phys.* **2019**, *52*, 075201. [CrossRef]
24. Madison, D.H.; Calhoun, R.V.; Shelton, W.N. Triple-differential cross sections for electron-impact ionization of helium. *Phys. Rev. A* **1977**, *16*, 552. [CrossRef]
25. De Lucio, O.; DuBois, R. Differential studies and projectile charge effects in ionization of molecular nitrogen by positron and electron impact. *Phys. Rev. A* **2016**, *93*, 032710. [CrossRef]
26. Avaldi, L.; Camilloni, R.; Fainelli, E.; Stefani, G. Ionization of the N_2 3 sigma g orbital by electron impact studied by asymmetric (e, 2e) experiments. *J. Phys. B At. Mol. Opt. Phys.* **1992**, *25*, 3551. [CrossRef]
27. Purohit, G.; Kato, D. Projectile charge effects on the differential cross sections for the ionization of molecular nitrogen by positrons and electrons. *J. Phys. B At. Mol. Opt. Phys.* **2018**, *51*, 135202. [CrossRef]

28. Tóth, I.; Campeanu, R.; Chiş, V.; Nagy, L. Electron impact ionization of diatomic molecules. *Eur. Phys. J. D* **2008**, *48*, 351–354. [CrossRef]
29. Lahmam-Bennani, A.; Casagrande, E.S.; Naja, A. Experimental investigation of the triple differential cross section for electron impact ionization of N_2 and CO_2 molecules at intermediate impact energy and large ion recoil momentum. *J. Phys. B At. Mol. Opt. Phys.* **2009**, *42*, 235205. [CrossRef]
30. Senftleben, A.; Al-Hagan, O.; Pflüger, T.; Ren, X.; Madison, D.; Dorn, A.; Ullrich, J. Fivefold differential cross sections for ground-state ionization of aligned H_2 by electron impact. *J. Chem. Phys.* **2010**, *133*, 044302. [CrossRef]
31. Madison, D.H.; Al-Hagan, O. The distorted-wave Born approach for calculating electron-impact ionization of molecules. *J. At. Mol. Phys.* **2010**, *2010*. [CrossRef]
32. Lee, C.; Yang, W.; Parr, R.G. Development of the Colle-Salvetti correlation-energy formula into a functional of the electron density. *Phys. Rev. B* **1988**, *37*, 785. [CrossRef] [PubMed]
33. Furness, J.; McCarthy, I. Semiphenomenological optical model for electron scattering on atoms. *J. Phys. B At. Mol. Phys.* **1973**, *6*, 2280. [CrossRef]
34. Riley, M.E.; Truhlar, D.G. Approximations for the exchange potential in electron scattering. *J. Chem. Phys.* **1975**, *63*, 2182–2191. [CrossRef]
35. Padial, N.; Norcross, D. Parameter-free model of the correlation-polarization potential for electron-molecule collisions. *Phys. Rev. A* **1984**, *29*, 1742. [CrossRef]
36. Ward, S.; Macek, J. Wave functions for continuum states of charged fragments. *Phys. Rev. A* **1994**, *49*, 1049. [CrossRef]

MDPI
St. Alban-Anlage 66
4052 Basel
Switzerland
Tel. +41 61 683 77 34
Fax +41 61 302 89 18
www.mdpi.com

Atoms Editorial Office
E-mail: atoms@mdpi.com
www.mdpi.com/journal/atoms

www.ingramcontent.com/pod-product-compliance
Lightning Source LLC
Chambersburg PA
CBHW062049220926
51850CB00010B/2708
9783036570112